Praise for *The Retro Future*

As John Michael Greer writes in *The Retro Future: Looking to the Past to Reinvent the Future,* when you've driven down a blind alley and are sitting there with your bumper pressed against a brick wall, the only way forward is by backing up... staying stuck against the brick wall leads nowhere useful. His call is not to a tribal or Stone Age existence, but rather, a return to things that actually work. Greer calls for "retrovation"—that is "retro plus innovation" rather than forsaking the old because we assume that all things new are superior. It's time to redefine "progress" in this way rather than using the latest technology to dig our heels into the illusion that "nothing is wrong at all." Engaging, witty, and exemplary of the Greer style we've come to love and rely on, *The Retro Future* will not disappoint. In fact, it may reassure you that in many instances, your ancestors got it right.

—Carolyn Baker, Ph.D., author, *Love In The Age of Ecological Apocalypse* and *Dark Gold: The Human Shadow And The Global Crisis.*

Whether or not you accept John Michael Greer's argument that a deindustrialized future is inevitable, you'll appreciate his call for the freedom to select the best technologies of the past—worthy and sustainable tools, not pernicious prosthetics. Greer's vision of a "post-progress" world is clear, smart, and ultimately hopeful.

—Richard Polt, professor of philosophy, Xavier University; author, *The Typewriter Revolution: A Typist's Companion for the 21st Century*

What might your life be like without an automobile, TV, or a mobile phone? Ask John Michael Greer, who lives that way and recommends it as practice for the soon-to-be-normal. Greer says we are embarked upon the post-progress era. Climate change, loose nukes, and resource exhaustion are among its many challenges. In *The Retro Future,* Greer looks backward to mark the way forward.

—Albert Bates, author, *The Post-Petroleum Survival Guide, The Biochar Solution,* and *The Paris Agreement.*

The prevailing assumption is that you will accept every bit of new technology, whether enthusiastically or grudgingly, or you won't be able to spend your life in a box, tethered to a gadget, looking at colored pixels. Greer's book discards this assumption: it is up to you to order your technology à la carte, plus the box and the gadget are soon going away in any case. And the stunning bit of news is, most of the recent technological progress has been in the direction of shoddy, inconvenient, short-lived, buggy time-wasters, so there is a lot for you to reject.

—Dmitry Orlov, author, *Reinventing Collapse* and *Five Stages of Collapse*

THE RETRO FUTURE

{ LOOKING TO THE PAST TO REINVENT THE FUTURE }

JOHN MICHAEL GREER

new society
PUBLISHERS

Cover design by Diane McIntosh.
Glasses © iStock (523394612); night sky © iStock (531473846).
Numbers © Vrender; initial caps © jro-grafik; p. 1 © lisaalisa_ill / Adobe Stock.

Printed in Canada. First printing September, 2017.

Inquiries regarding requests to reprint all or part of *The Retro Future* should be addressed to New Society Publishers at the address below.
To order directly from the publishers, please call toll-free (North America) 1-800-567-6772, or order online at www.newsociety.com

Any other inquiries can be directed by mail to:

New Society Publishers
P.O. Box 189, Gabriola Island, BC V0R 1X0, Canada
(250) 247-9737

LIBRARY AND ARCHIVES CANADA CATALOGUING IN PUBLICATION

Greer, John Michael, author
The retro future : looking to the past to reinvent the future / John Michael Greer.

Includes bibliographical references and index.
Issued in print and electronic formats.
ISBN 978-0-86571-866-1 (softcover).—ISBN 978-1-55092-658-3 (PDF).—
ISBN 978-1-77142-253-6 (EPUB)

1. Progress. 2. Regression (Civilization). 3. Civilization, Modern—21st century—Forecasting. I. Title.

HM891.G745 2017 303.44 C2017-903762-5
C2017-903763-3

Funded by the Government of Canada | Financé par le gouvernement du Canada | Canadä

New Society Publishers' mission is to publish books that contribute in fundamental ways to building an ecologically sustainable and just society, and to do so with the least possible impact on the environment, in a manner that models this vision.

Contents

Preface

In a certain sense, this book is part of a trilogy, though it differs from most trilogies in that the books can be read in any order. The books in question came into being out of a growing sense on my part that the predicament of our time could not be understood from within the conventional wisdom that created it, and that the most important element of that conventional wisdom—the heart of a secular belief system that shares most of the characteristics of a religion—was faith in progress.

My first explorations of that theme focused on understanding where the ersatz religion of progress came from and how the mismatch between faith in progress and the insistent reality of our society's failure to progress—or, put more forcefully, of the opening stages of its decline—was likely to play out in the thought, imagination, and beliefs of people in the contemporary world. Those explorations eventually gave rise to a book, *After Progress: Religion and Reason at the End of the Industrial Age.*[1] As that first reconnaissance reached clarity, I began two other related projects, both oriented toward figuring out what sorts of responses might be appropriate to the end of the age of progress.

One of those projects used narrative fiction to try to explore the prospects of a society that abandoned the religion of perpetual progress and, instead, allowed itself and its citizens to pick and choose among the technologies and lifestyles already explored by our species. That narrative became a novel, *Retrotopia.*[2] The other project approached the same question from the more conventional angle of nonfiction, and the result is the book you are holding in your hands right now.

As discussed later in this book, the idea of an end to progress is freighted with a great many irrational terrors and strange beliefs. It's far from uncommon for people to insist that any future that isn't defined by the endless elaboration of already overelaborate technologies must somehow involve going back to the caves or sinking into medieval squalor or being gobbled up by any of the other hobgoblins of the past with which the religion of progress threatens unbelievers. These reactions have deep emotional roots; for several centuries now, a vast number of people in the industrial world have allowed their sense of meaning, purpose, and value to depend on their assumed role in the grand onward march of progress from the caves to the stars, and letting go of that self-image is a very challenging thing.

That said, it's not as though we ultimately have a choice. On the one hand, the exhaustion of nonrenewable resources and the buildup of pollutants in the atmosphere, the seas, and the soil are already starting to impose a rising spiral of costs on further attempts to make our technologies even more elaborate than they are today. On the other, it's becoming increasingly clear to people in the industrial world that progress does not necessarily mean improvement, and that older and simpler technologies very often do a better job at their tasks than the latest hypercomplex, high-tech equivalent. A growing number of people are thus beginning to turn aside from the products of progress. That these older, simpler technologies are very often less dependent on nonrenewable resources and less damaging to the biosphere that supports all our lives is just one benefit of that heretical but necessary act.

This book seeks to discuss what the world looks like in the wake of the end of progress: why progress is ending, why it could never have fulfilled the overblown promises made in its name, and what the prospects of our society and species might look like as the age of progress gives way to an age of environmental blowback and technological unraveling. It's popular to paint those latter prospects in unremittingly bleak colors, but here again that reflects the unthinking assumptions of our age rather than the facts as they actually exist. The burdens that progress have piled upon us, as individuals, as communities, and as a species, are not small, and once the shock has passed off, liberation from those burdens may well be experienced by many of us as a reason for celebration rather than mourning.

That said, there are serious downsides to the end of progress, just as there were equally serious downsides to its beginning and to every step of its historical course. My hope is that this book, as a first survey of the almost entirely unexplored landscape on the far side of progress, will help my readers prepare themselves for the largely unexpected future ahead of us.

My previous books have had a variety of intellectual debts, but this one has depended almost entirely on one source—the readership of my former blog "The Archdruid Report." For eleven years, from the first tentative posts about peak oil and the future of industrial society all the way to the last posts about the nature of human experience, my readers encouraged me, argued with me, brought me data points that confirmed or challenged the ideas that I've offered, and in general created a congenial and thought-provoking environment for the development of my ideas. My thanks go to all.

1

THE END OF PROGRESS

MOST PEOPLE in the industrial world believe that the future is, by definition, supposed to be better than the past, that growth is normal and contraction is not, that newer technologies are superior to older ones, and that the replacement of simple technologies by complex ones is as unstoppable as it is beneficent. That's the bedrock of the contemporary faith in progress.[1] This faith remains unchallenged by most people today, even though the evidence of our everyday lives contradicts it at every turn.

Most of us know perfectly well that every software "upgrade" these days has more bugs and fewer useful features than what it replaced, and every round of "new and improved" products hawked by the media and shoveled onto store shelves is more shoddily made, more loaded with unwanted side effects, and less satisfactory at meeting human needs than the last round. Somehow, though, a good many of the people who witness this reality, day in and day out, still manage to insist that the future will be, or at least ought to be, a paradise propped up by perfectly functioning machines, in which all the latest clichés about the future will inevitably come true. That the rising tide of technological failure might be something other than an accidental roadbump on the way to utopia—that it might be trying to tell us something that, by and large, we don't want to hear—has not yet entered our society's darkest dream.

Meanwhile, as problems mount and solutions run short, the contemporary faith in progress drives a common insistence that it's never too late to save the world. No matter how troubling the signs on the horizon, no matter how many predictions of impending trouble have turned into descriptions of troubles we're facing here and now, it's astonishingly rare for anyone to notice that we're past the point where it makes any sense to sit around talking about how somebody ought to fix things one of these days.

The events of our time, though, show no particular interest in waiting until we get around to dealing with them. At least three factors at work in today's world—peak oil, and more generally the peaking of global production of fossil fuels; the ongoing failure of alternative energy technologies to replace fossil fuels; and the accelerating pace of anthropogenic climate change—are already having a major impact on the global economy and, increasingly, on other aspects of human and nonhuman life as well.

Those issues could have been faced and dealt with as soon as it became clear that they were going to be problematic. In every case, there were straightforward fixes available, and if they had been put into place as soon as the facts showed that trouble was on its way, the necessary changes could have been made gradually, without overturning the whole structure of society. But that's not what happened. Instead, obsolete policies stayed frozen in place while the opportunities for constructive change slipped past. Now the bill is coming due.

This doesn't mean that action is useless, much less that we should huddle down, close our eyes, and wait for the end. It means, rather, that business as usual will not last much longer, no matter what we do about it. In the decades ahead, many things that people in the industrial world consider normal will go away forever. That's going to be profoundly difficult, but it's also profoundly liberating, because the struggle to maintain the status quo has been a massive force blocking the way to constructive change. As the familiar landscapes of the industrial age give way to the unexpected vistas of the near and middle future, the focus of meaningful action will have to shift from preservation to remediation, from "How can we keep our familiar ways of doing things?" to "Now that the familiar things are gone, what can we put in their place?"

The Law of Diminishing Returns

I'm well aware that asking people in the early twenty-first century to doubt the omnipotence and eternal goodness of progress ranks right up there with suggesting to a medieval peasant that God and his saints and angels aren't up there in heaven any more. There are nonetheless two crucial reasons why cumulative technological progress, of the sort that's reshaped the industrial world over the past three centuries, was a temporary, self-limiting process that often imposed costs that outweighed its benefits.

The first is the law of diminishing returns—the principle that the more often you repeat a given action, the fewer benefits you get from each successive repetition and the more the costs mount up. Nearly everything in the world of human experience is subject to this law. The process of extracting petroleum from the earth is a good example: the first oil wells made huge profits for very little expense, while the hunt for the last scrapings of the bottom of the planet's oil barrel that occupies the petroleum industry today has extraordinarily high costs and meager returns.

Is technological progress subject to the same principle? Believers in progress like to insist that this can't be the case, but the evidence suggests otherwise. Consider the way that energy technologies have become more and more expensive to develop over time. The steam engine, the first major energy technology innovation in modern times, was invented by working engineers in their off hours, using ordinary pipefitting tools. The internal combustion engine and the electrical generator required more systematic effort, but were still well within the reach of a single inventor working in a laboratory. Nuclear fission required an expenditure of money and resources so huge that only a handful of relatively rich nations could afford it. Commercial nuclear fusion power, as we'll see in Chapter Two, is turning out to be so costly that nobody anywhere can afford it at all.

In exactly the same way, and for many of the same reasons, the first advances in health care—basic sanitation, antiseptics, and vaccination—cost very little and brought immense benefits. With every passing year, costs went up and benefits went down, until current health care research is investing billions of dollars in projects that may benefit only a few people, if any. The low-hanging fruit got picked first, leaving more difficult projects for later.

The same is true in every other field. This is why, as a detailed study of patent records has shown, the modern world hit its peak of innovation in the last quarter of the nineteenth century, and the pace of technological progress has actually decreased steadily since that time.[2] This implies that, at some point, the benefits of continued technological progress will no longer equal the costs and progress will grind to a halt because it no longer pays for itself. Thus there can be such a thing as too much technology, and a very strong case can be made that in the world's industrial nations we've already gotten well past that point.

In a typically cogent article,[3] maverick economist Herman Daly sorted out the law of diminishing returns into three interacting processes. The first is *diminishing marginal utility*—that is, the more of anything you have, the less any additional increment of that thing contributes to your well-being. If you're hungry, one sandwich is a very good thing; two is pleasant; three is a luxury; and somewhere beyond that, when you've given sandwiches to all your coworkers, the local street people, and anyone else you can find, more sandwiches stop being any use to you. When more of anything no longer brings any additional benefit, you've reached the point of futility, at which further increments are a waste of time and resources.

Well before that happens, though, two other factors come into play. First, it costs you almost nothing to cope with one sandwich, and very little more to cope with two or three. After that you start having to invest time, and quite possibly resources, in dealing with all those sandwiches, and each additional sandwich adds to the total burden. Economists call that *increasing marginal disutility*—that is, the more of anything you have, the more any additional increment of that thing is going to cost you, in one way or another. Somewhere in there, too, there's the impact that dealing with those sandwiches has on your ability to deal with other things you need to do; that's the *increasing risk of whole-system disruption*—the more of anything you have, the more likely it is that an additional increment of that thing is going to disrupt the wider system in which you exist.

Next to nobody wants to talk about the way that technological progress has already passed the point of diminishing returns in all three senses: that the marginal utility of each new round of technology is dropping fast; the marginal disutility is rising at least as fast; and whole-system disruptions driven by technology are becoming an inescapable presence in everyday life. Still, an uncomfortable awareness of that fact is becoming increasingly common these days, however subliminal it may be, and is beginning to have a popular culture among many other things.

If you've dug yourself into a hole, as the saying goes, the first thing you need to do is stop digging. If a large and growing fraction of your society's problems are being caused by too much technology applied with too little caution, similarly, it's not exactly helpful to insist that applying even more technology with even less skepticism about its consequences is the only possible answer to those problems.

There's a useful word for something that remains stuck in a culture after the conditions that once made it relevant have passed away, and that word is "superstition." The beliefs that more technology is always better, that every problem must have a technological solution, and that technology always solves more problems than it creates, are among the prevailing superstitions of our time. I'd like to suggest that, comforting and soothing as those superstitions may be, it's high time we outgrow them and deal with the hard realities of a world in which taking such faith-based notions as a guide to the future may not be sensible, or even sane.

Yet there's another reason to ask hard questions about where progress is taking us, and it unfolds from the issue of externalities. Externalities are the costs of an economic activity that aren't paid by the buyer or the seller directly but are pushed off onto some third party. You won't hear much about externalities these days; it many circles, it's considered impolite to mention them, but they're a pervasive presence in contemporary life, and they play a very large role in some of the most intractable problems of our age. Some of those problems were discussed by Garrett Hardin in his famous essay on the tragedy of the commons, and more recently by Elinor Ostrom in her studies of how that tragedy can be avoided.[4] Still, I'm not sure how often it's recognized that the phenomena they discussed applies not just to commons but to societies as a whole—especially to societies like ours.

A simplified example may be useful here. Let's imagine a blivet factory, then, that turns out three-pronged blivets in pallet loads for customers. The blivet-making process, like all other manufacturing, produces waste as well as blivets, and we'll assume for the sake of the example that blivet waste is moderately toxic and causes health problems in people who ingest it. The blivet factory produces one barrel of blivet waste for every pallet of blivets it ships. The cheapest option for dealing with the waste, and thus the option that economists favor, is to dump it into the river that flows past the factory.

Notice what happens if the blivet manufacturer follows this approach. The manufacturer has maximized his own benefit from the manufacturing process by avoiding the expense of finding some other way to deal with all those barrels of blivet waste. His customers also benefit, because blivets cost less than they would if the cost of waste disposal was factored into the price. On the other hand, the costs of dealing with the blivet waste don't disappear; they are imposed on the people downstream who get their drinking water directly or indirectly from the river and who suffer from health problems because there's blivet waste in their water. The blivet manufacturer is thus externalizing the cost of waste disposal; his increased profits are being paid for at a remove by the increased health care costs of everyone downstream.

That's how externalities function. Back in the days when people actually talked about the downsides of economic growth, there was a lot of discussion of how to handle externalities, and not just on the leftward end of the spectrum. I recall a thoughtful book titled *TANSTAAFL*—that's an acronym, for those who don't know their Heinlein, for "There Ain't No Such Thing As A Free Lunch"[5]—which argued, on solid libertarian-conservative grounds, that the environment could best be preserved by making sure that everyone paid full sticker price for the externalities they generated. Today's crop of American pseudo-conservatives, of course, turned their back on all this a long time ago and insist at the top of their lungs on their allegedly God-given right to externalize as many costs as they possibly can. This is all the more ironic in that most pseudo-conservatives claim to worship a God who said some very specific things about "what ye do unto the least of these," but that's a subject for a different time.

The Externality Trap

Economic life in the industrial world these days can be described, without too much inaccuracy, as an arrangement set up to allow a privileged minority to externalize nearly all their costs onto the rest of society while pocketing as many as possible of the benefits themselves. That's come in for a certain amount of discussion in recent years,[6] but I'm not sure how many of the people who've participated in those discussions have given any thought to the role that technological progress plays in facilitating the internalization of benefits and the externalization of costs that drive today's increasingly inegalitarian societies. Here again, our example will be helpful.

Before the invention of blivet-making machinery, let's say, blivets were made by old-fashioned blivet makers, who hammered them out on iron blivet anvils in shops that were to be found in every town and village. Like other handicrafts, blivet-making was a living rather than a ticket to wealth; blivet makers invested their own time and muscular effort in their craft and turned out enough in the way of blivets to meet the demand. Notice also the effect on the production of blivet waste. Since blivets were being made one at a time rather than in pallet loads, the total amount of waste was smaller; the conditions of handicraft production also meant that blivet makers and their families were more likely to be exposed to the blivet waste than anyone else, and so they had an incentive to invest the extra effort and expense to dispose of it properly. Since blivet makers were ordinary craftspeople rather than millionaires, furthermore, they weren't as able to buy exemption from local health laws.

The invention of the mechanical blivet press changed that picture completely. Since one blivet press could do as much work as fifty blivet makers, the income that would have gone to those fifty blivet makers and their families went instead to one factory owner and his stockholders, with as small a share as possible set aside for the wage laborers who operated the blivet press. The factory owner and stockholders had no incentive to pay for the proper disposal of the blivet waste, either—quite the contrary, since having to meet the disposal costs cut into their profit, buying off local governments was much cheaper, and, if the harmful effects of blivet waste were known, you can bet that the owner and shareholders all lived well upstream from the factory.

Notice also that a blivet manufacturer who paid a living wage to his workers and covered the costs of proper waste disposal would have to charge a higher price for blivets than one who did neither and thus would be driven out of business by his more ruthless competitor. Externalities aren't simply made possible by technological progress, in other words; they're the inevitable result of technological progress in a market economy, because the more a firm externalizes the costs of production, the more readily it can outcompete rival firms, and the firm that succeeds in externalizing the largest share of its costs is the most likely to survive and prosper.

Each further step in the progress of blivet manufacturing, in turn, tightened the same screw another turn. Today, to finish up the metaphor, the entire global supply of blivets is made in a dozen factories in distant Slobbovia, where sweatshop labor under ghastly working conditions and the utter absence of environmental regulations make the business of blivet fabrication more profitable than anywhere else. The blivets are as shoddily made as possible; the entire blivet supply chain, from the open-pit mines worked by slave labor that provide the raw materials to the big box stores with part-time, poorly paid staff selling blivetronic technology to the masses, is a human and environmental disaster. Every possible cost has been externalized, so that the two multinational corporations that dominate the global blivet industry can maintain their profit margins and pay absurdly high salaries to their CEOs.

That in itself is bad enough, but let's broaden the focus to include the whole systems in which blivet fabrication takes place: the economy as a whole, society as a whole, and the biosphere as a whole. The impact of technology on blivet fabrication in a market economy has predictable and well-understood consequences for each of these whole systems, which can be summed up precisely in the language we've already used. In order to maximize its own profitability and return on shareholder investment, the blivet industry externalizes costs in every available direction.

Since nobody else wants to bear those costs, either, most of them end up being passed onto the whole systems just named, because the economy, society, and the biosphere have no voice in today's economic decisions.

Like the costs of dealing with blivet waste, though, the other externalized costs of blivet manufacture don't go away just because they're externalized. As externalities increase, they tend to degrade the whole systems onto which they're dumped—the economy, society, and the biosphere. This is where the trap closes tight, because blivet manufacturing exists within those whole systems and can't be carried out unless all three systems are sufficiently intact to function in their usual way. As those systems degrade, their ability to function degrades also, and eventually one or more of them breaks down—the economy plunges into a depression; the society disintegrates into anarchy or totalitarianism; the biosphere shifts abruptly into a new mode that lacks adequate rainfall for crops—and the manufacture of blivets stops because the whole system that once supported it has stopped doing so.

Notice how this works out from the perspective of someone who's benefiting from the externalization of costs by the blivet industry—the executives and stockholders in a blivet corporation, let's say. As far as they're concerned, until very late in the process, everything is fine and dandy: each new round of technological improvements in blivet fabrication increases their profits, and if each such step in the onward march of progress also means that economies go haywire, democratic institutions implode, toxic waste builds up in the food chain, or what have you, hey, that's not their problem—and after all, that's just the normal, praiseworthy creative destruction of capitalism, right?

That sort of insouciance is easy for at least three reasons. First, the impacts of externalities on whole systems can pop up a very long way from the blivet factories. Second, in a market economy, everyone else is externalizing their costs as enthusiastically as the blivet industry, and so it's easy for blivet manufacturers (and everyone else) to insist that whatever's going wrong is not their fault. Third, and most crucially, whole systems as stable and enduring as economies, societies, and biospheres can absorb a lot of damage before they tip over into instability. The process of externalization of costs can thus run for a very long time, and become entrenched as a basic economic habit, long before it becomes clear to anyone that continuing along the same route is a recipe for disaster.

Even when externalized costs have begun to take a visible toll on the economy, society, and the biosphere, furthermore, any attempt to reverse course faces nearly insurmountable obstacles. Those who profit from the existing order of things can be counted on to fight tooth and nail for the right to keep externalizing their costs: after all, they have to pay the full price for any reduction in their ability to externalize costs, while the benefits created by not imposing those costs on whole systems are shared among all participants in the economy, society, and the biosphere respectively. Nor is it necessarily easy to trace back the causes of any given whole-system disruption to specific externalities benefiting specific people or industries. It's rather like loading hanging weights onto a chain; sooner or later, as the amount of weight hung on the chain goes up, the chain is going to break, but the link that breaks may be far from the last weight that pushed things over the edge, and every other weight on the chain made its own contribution to the end result.

A society that's approaching collapse because too many externalized costs have been loaded onto the whole systems that support it thus shows certain highly distinctive symptoms. Things are going wrong with the economy, society, and the biosphere, but nobody seems to be able to figure out why; the measurements that economists use to determine prosperity show contradictory results, with those that measure the profitability of individual corporations and industries giving much better readings than those that measure the performance of whole systems; the rich are convinced that everything is fine, while outside the narrowing circles of wealth and privilege, people talk in low voices about the rising spiral of problems that beset them from every side. If this doesn't sound familiar to you, dear reader, you probably need to get out more.

At this point it may be helpful to sum up the argument I've developed here:

a) Every increase in technological complexity tends also to increase the opportunities for externalizing the costs of economic activity.
b) Market forces make the externalization of costs mandatory rather than optional, since economic actors that fail to externalize costs will tend to be outcompeted by those that do.
c) In a market economy, as all economic actors attempt to externalize as many costs as possible, externalized costs will tend to be passed on preferentially and progressively to whole systems such as the economy, society, and the biosphere, which provide necessary support for economic activity but have no voice in economic decisions.
d) Given unlimited increases in technological complexity, there is no necessary limit to the loading of externalized costs onto whole systems, short of systemic collapse.
e) Unlimited increases in technological complexity in a market economy thus necessarily lead to the progressive degradation of the whole systems that support economic activity.
f) Technological progress in a market economy is therefore self-terminating, and ends in collapse.

Secular Stagnation

There are, of course, arguments that could be deployed against this thesis. For example, it could be argued that progress doesn't have to generate a rising tide of externalities. The difficulty with this argument is that externalization of costs isn't an accidental side effect of technology but an essential aspect—it's not a bug, it's a feature. Every technology is a means of externalizing some cost that would otherwise be borne by a human body. Even something as simple as a hammer takes the wear and tear that would otherwise affect the heel of your hand, let's say, and transfers it to something else: directly, to the hammer; indirectly, to the biosphere by way of the trees that had to be cut down to make the charcoal to smelt the iron, the plants that were shoveled aside to get the ore, and so on.

The more complex a technology becomes, the more costs it generates, since every bit of added complexity has to be paid for. Each more-complex technology thus has to externalize its additional costs in order to compete against the simpler technology it replaces. In the case of a hypercomplex technosystem such as the internet, the process of externalizing costs has gone so far, through so many tangled interrelationships, that it's next to impossible to figure out exactly who's paying for how much of the gargantuan inputs needed to keep the thing running. This lack of transparency feeds the illusion that large systems are cheaper than small ones, by making externalities of scale look like economies of scale.

It might be argued instead that a sufficiently stringent regulatory environment, forcing economic actors to absorb all the costs of their activities instead of externalizing them onto others, would be able to stop the degradation of whole systems while still allowing technological progress to continue. The difficulty here is that increased externalization of costs is what makes progress profitable. All other things being equal, a complex technology will be more expensive in real terms than a simpler technology, for the simple fact that each additional increment of complexity has to be paid for by an investment of energy and other forms of real capital.

Strip complex technologies of the subsidies that transfer some of their costs to the government, the perverse regulations that transfer some of their costs to the rest of the economy, the bad habits of environmental abuse and neglect that transfer some of their costs to the biosphere, and so on, and pretty soon you're looking at hard economic limits to technological complexity, as people forced to pay the full sticker price for complex technologies maximize the benefits they receive by choosing simpler, more affordable options instead. A regulatory environment sufficiently strict to keep technology from accelerating to collapse would thus bring technological progress to a halt by making it unprofitable.

Notice, however, the flipside of the same argument: a society that chose to stop progressing technologically could maintain itself indefinitely, so long as its technologies weren't dependent on nonrenewable resources or the like. The costs imposed by a stable technology on the economy, society, and the biosphere would be more or less stable, rather than increasing over time, and it would therefore be much easier to figure out how to balance out the negative effects of those externalities and maintain the whole system in a steady state. Societies that treated technological progress as an option rather than a requirement, and recognized the downsides to increasing complexity, could also choose to reduce complexity in one area in order to increase it in another, and so on—or they could just raise a monument to the age of progress and go do something else instead.

The costs of progress are already starting to take an increasing bite out of the global economy. That bite shows up in the fact that none of the twenty biggest industries in today's world could break even, much less make a profit, if they had to pay for the damage they do to the environment.[7] The conventional wisdom these days is that it's unfair to make those industries pay for the costs they impose on the rest of us. That attitude is exemplified by fracking firms in North Dakota, among many others, who proposed at height of the fracking bubble that they should be exempted from rules for handling radioactive waste from their wells, because following the rules would prevent them from making a profit.[8] That the costs externalized by the fracking industry will sooner or later be paid by others, as radionuclides in fracking waste begin to generate cancer clusters, is not something that our current economic thought is able to grasp.

The crucial point to keep in mind is that externalized costs don't just go away. They will be paid by someone; the only question is who pays them. That's the central argument of *The Limits to Growth*, still the most accurate (and thus inevitably the most reviled) of the studies that tried unavailingly to guide industrial society away from self-inflicted ruin: on a finite planet, once an inflection point is passed, the costs of economic growth rise faster than growth does, and sooner or later force the global economy to its knees.[9] The mere fact that those costs aren't carried on the balance sheets of the companies that generate them doesn't make those costs go away; it just keeps them from being taken into account by policymakers.

One way in which those costs may already be having an impact is in the phenomenon of secular stagnation.[10] It so happens that when you subtract the paper wealth manufactured by derivatives and other forms of financial make-believe, the global economy has been stuck in a period of slow, no, or negative growth since 2009. There are plenty of economists, mind you, who insist that this can't happen, and even among those who admit that what's happening can indeed happen, there's no consensus as to how or why such a thing could occur. I'd like to suggest that the most important cause of secular stagnation is the increasing impact of externalities on the economy. The dubious bookkeeping that leads economists to think that externalized costs go away because they're not entered into anyone's books doesn't actually make them disappear, after all. Instead, they become an unrecognized burden on the economy as a whole, an unfelt headwind blowing with hurricane force in the face of economic growth.

Thus the insistence by fracking firms that they ought to be allowed to externalize even more of their costs in order to maintain their profit margin is self-defeating, even if the firms themselves don't realize that. If in fact the buildup of externalized costs is what's causing the ongoing economic slowdown that's driving down commodity prices, forcing down interest rates, and resurrecting the specter of deflationary depression, the fracking firms in question are trying to respond to secular stagnation by doing more of what causes secular stagnation.

In theory, this sort of self-defeating behavior would be recognized for what it is and set aside as a bad idea. In the real world, by contrast, fracking firms, like every other business concern these days, can be expected to put their short-term cash flow ahead of the survival of their industry, or for that matter of industrial civilization as a whole. That's business as usual—and it's made even easier than it otherwise would be by certain habits of thought that make it hard to think clearly about technology and progress.

Technologies in the Plural

Here's an example. When talking heads these days babble about technology in the singular, as a uniform, monolithic thing that progresses according to some relentless internal logic of its own, they're spouting balderdash.[11] In the real world, there's no such monolith, no technology in the singular. Instead, there are technologies in the plural, clustered more or less loosely in technological suites that may or may not have any direct relation to one another.

An example, again, might be useful here. Consider the technologies necessary to build a steel-framed bicycle. The metal parts require the particular suite of technologies we use to smelt ores, combine the resulting metals into useful alloys, and machine and weld those into shapes that fit together to make a bicycle. The tires, inner tubes, brake pads, seat cushion, handlebar grips, and paint require a different suite of technologies drawing on various branches of applied organic chemistry, and a few other suites also have a place: for example, the one that's needed to make and apply lubricants.

The suites that make a bicycle have other uses. If you can build a bicycle, as Orville and Wilbur Wright demonstrated, you can also build an aircraft and a variety of other interesting machines as well. That said, there are other technologies—say, the ones needed to manufacture medicines or precision optics or electronics—that require very different technological suites. You can have everything you need to build a bicycle and still be unable to make a telescope or a radio receiver, and vice versa.

Strictly speaking, therefore, nothing requires all the different technological suites to move in lockstep. It would have been quite possible for different technological suites to have appeared in a different order than they did; it would have been just as possible for some of the suites central to our technologies today to have never gotten off the ground, while other technologies we never tried emerged instead. Imagine, for example, an alternative reality in which solar water heaters (in our world, worked out by 1920) and passive solar architecture (mostly developed in the 1960s and 1970s) were standard household features, canal boats (dating from before 1800) and tall ships (ditto) were the primary means of bulk transport, shortwave radio (developed in the early twentieth century) was the standard long-range communications medium, ultralight aircraft (largely developed in the 1980s) were still in use, and engineers crunched numbers using slide rules (perfected around 1880).

There's no reason why such a pastiche of technologies from different eras couldn't work. We know this because what passes for modern technology is a pastiche of the same kind, in which (for example) cars whose basic design dates from the 1890s are gussied up with computers invented a century later. Much of modern technology, in fact, is old technology with a new coat of paint and a few electronic gimmicks tacked on, and it's old technology that originated in many different eras, too. Part of what differentiates modern technology from older equivalents, in other words, is mere fashion. Another part, though, moves into more explosive territory.

One reader of my blog "The Archdruid Report" once enlivened the discussion on the comments page with the story of the one and only class on advertising she took at college. The teacher invited a well-known advertising executive to come and talk about the business, and one of the points he brought up was the marketing of disposable razors. The old-fashioned steel safety razor, the guy admitted cheerfully, was a much better product: it was more durable, less expensive, and gave a better shave than disposable razors. Unfortunately, it didn't make the kind of profits for the razor industry that the latter wanted, and so the job of the advertising company was to convince shavers that they really wanted to spend more money on a worse product instead.

I know it may startle some people to hear a luxuriantly bearded man talk about shaving, but I do have a certain amount of experience with the process—though admittedly it's been many years. The executive was quite correct: an old-fashioned safety razor with interchangeable blades gives better shaves than a disposable. What's more, an old-fashioned safety razor combined with a shaving brush, a cake of shaving soap, a mug, and a bit of hot water from the teakettle produces a shaving experience that's vastly better, in every sense, than what you'll get from squirting chemical-laced foam out of a disposable can and then scraping your face with a disposable razor; it takes the same amount of time, costs much less on a per-shave basis, and has a drastically smaller ecological footprint to boot.

Notice also the difference in the scale and complexity of the technological suites needed to maintain these two ways of shaving. To shave with a safety razor and shaving soap, you need the metallurgical suite that produces razors, the very simple household-chemistry suite that produces soap, the ability to make pottery and brushes, and some way to heat water. To shave with a disposable razor and a can of squirt-on shaving foam, you need fossil fuels for plastic feedstocks, chemical plants to manufacture the plastic and the foam, the whole range of technologies needed to manufacture and fill the pressurized can, and so on—all so that you can count on getting an inferior shave at a higher price, and the razor industry can boost its quarterly profits.

That's a small, and arguably silly, example of a vast and far from silly issue. These days, when you see the words "new and improved" on a product, rather more often than not, the only thing that's been improved is the bottom line of the company that's trying to sell it to you. When you hear equivalent claims about some technology that's being marketed to society as a whole, rather than sold to you personally, the same rule applies at least as often.

What, after all, defines a change as "progress"? There's a wilderness of ambiguities, some of them quite deliberate, hidden in that apparently simple word. The contemporary faith in progress presupposes that there's an inherent dynamic to history and that things therefore change, or tend to change, or at the very least ought to change, from worse to better over time. That presupposition then gets flipped around into the even more dubious claim that just because something's new, it must be better than whatever it replaced. Move from there to specific examples, and all of a sudden it's necessary to deal with competing claims—if there are two hot new technologies on the market, is option A more progressive than option B, or vice versa? The answer, of course, is that whichever of them manages to elbow the other aside, by whatever means, will be retroactively awarded the coveted title of the next step in the march of progress.

That was exactly the process by which the appropriate tech of the 1970s was shoved aside and buried in the memory hole of our culture.[12] In its heyday, appropriate tech was as cutting edge as anything you care to name, a rapidly advancing field pushed forward by brilliant young engineers and innovative startups, and it saw itself (and presented itself to the world) as the wave of the future. In the wake of the Reagan-Thatcher counterrevolution of the 1980s, though, it was retroactively stripped of its progressive label and consigned to the dustbin of the past. Technologies that had been lauded in the media as brilliantly innovative in 1978 were thus being condemned in the same media as Luddite throwbacks by 1988. If that abrupt act of redefinition reminds any of my readers of the way history got rewritten in George Orwell's *1984*—"Oceania has never been allied with Eurasia" and the like—well, let's just say the parallel was noticed at the time, too.

The same process on a much smaller scale can be traced with equal clarity in the replacement of the safety razor and shaving soap with the disposable razor and squirt-can shaving foam. In what sense is the latter, which wastes more resources and generates more trash in the process of giving users a worse shave at a higher price, an advance (that metaphor again) on the former? Merely the fact that it's been awarded that title by advertising and the media. If razor companies could make more money by reintroducing the Roman habit of scraping beard hairs off the face with a chunk of pumice, no doubt that would quickly be proclaimed as the last word in cutting-edge, up-to-date hypermodernity, too.

What Progress Actually Means

Behind the mythological image of the relentless and inevitable forward march of technology-in-the-singular in the grand cause of progress, in other words, lies a murky underworld of crass commercial motives and no-holds-barred struggles over which of the available technologies will get the funding and marketing that will define it as the next step in the march of progress. That's as true of major technological programs as it is of shaving supplies. Some of my readers are old enough, as I am, to remember when supersonic airliners and undersea habitats were the next great steps in progress, until all of a sudden they weren't, and we may not be all that far from the point at which space travel and nuclear power will go the way of Sealab and the Concorde.

In today's industrial societies, we don't talk about that. It's practically taboo these days to mention the long, long list of waves of the future that abruptly stalled and rolled back out to sea without delivering on their promoters' overblown promises. Remind people that the same rhetoric currently being used to prop up faith in space travel, fusion power, or some other grand technological project was lavished just as thickly on these earlier failures, and you can expect to have that comment shouted down as an irrelevancy, if the other people in the conversation don't simply turn their backs and pretend that they never heard you say anything at all.

They have to do something of the sort, because the alternative is to admit that what we call "progress" isn't the impersonal, unstoppable force of nature that industrial culture's ideology insists it must be. Pay attention to the grand technological projects that failed, compare them with those that are failing now, and it's impossible to keep ignoring certain crucial if hugely unpopular points. To begin with, technological progress is a function of collective choices—do we fund Sealab or the Apollo program? Supersonic transports or urban light rail? Energy conservation and appropriate technology or an endless series of wars in the Middle East? No impersonal force makes those decisions; individuals and institutions make them, and then use the rhetoric of impersonal progress to cloak the political and financial agendas that guide the decision-making process.

What's more, even if the industrial world chooses to invest its resources in a project, the laws of physics and economics, not human preferences, determine whether the project is going to work. The Concorde is the poster child here, a white elephant that could never even cover its own operating costs. Like nuclear power, it was a technological success but an economic flop, only viable given huge and continuing government subsidies, and since the strategic benefits Britain and France got from having Concordes in the air were nothing like so great as those they got from having an independent source of raw material for nuclear weapons, it's not hard to see why the subsidies went where they did.

That is to say, when something is being lauded as the next great step forward in the glorious march of progress leading humanity to a better world someday, those who haven't drunk themselves tipsy on industrial civilization's folk mythology need to keep three things in mind. The first is that the next great step (etc.) might not actually work when it's brought down out of the billowing clouds of overheated rhetoric into the cold hard world of everyday life. The second is that even if it does work, the next great step (etc.) may be an inferior product, and do a less effective job of meeting human needs than whatever it's supposed to replace. The third is that when it comes right down to it, to label something as the next great step (etc.) is just a sales pitch, an overblown and increasingly trite way of saying "You really ought to buy this."

That implies, in turn, that it's entirely reasonable to respond by saying, "No, I like the thing I'm already using better"—or even to utter the unmentionable and say, "No, I'm going to use this other technology from the past because it works better." Get past the thoughtstopping rhetoric that insists you can't turn back the clock—of course you can; most of us do it every autumn without a second thought when daylight saving time ends—and it becomes hard not to notice that "progress" is just a label for whatever choices happen to have been made by governments and corporations, with or without input from the rest of us, and that if we don't like the choices that have been made for us in the name of progress, we can choose something else.

Of course, it's possible to stuff that sort of thinking back into the straitjacket of progress and claim that progress is chugging along just fine, and all we have to do is get it back on the proper track or what have you. This is a very common sort of argument and one that's been used over and over again by critics of this or that candidate for the next great step (etc.). The problem with that argument is that it may occasionally win battles but it consistently loses the war. By failing to challenge the folk mythology of progress and the unstated agendas that are enshrined by that mythology, it guarantees that no matter what technology or policy or program gets put into place, it'll end up leading to the same place as all the others before it.

That's the trap hardwired into the contemporary faith in progress. Once you buy into the notion that the specific choices made by industrial societies over the past three centuries or so are more than the projects that happened to win out in the struggle for wealth and power, once you let yourself believe that there's a teleology to it all—that is, that there's some objective goal called "progress" that these choices do a better or worse job of furthering—you've just made it much harder to ask the hard but necessary questions about where this thing called "progress" is going. The word "progress," remember, means going further in the same direction, and it's precisely questions about the direction that industrial society is going that most need to be asked.

I'd like to suggest that going further in the direction we've been going isn't a particularly bright idea just now. Going further in the direction we've been going means trying to expand per capita energy consumption in an era when fossil fuel reserves are depleting fast and the global economy is creaking and shuddering under the burden of increasingly costly fuel extraction. It means dumping ever more waste into the biosphere when the consequences of previous dumping are already bidding fair to threaten the survival of entire nations. On a less global scale, it also means shoddier products with louder advertising in a race to the bottom that's already gone very far.

Look at a trend that affects your life right now, and extrapolate it out in a straight line; that's what going further in the same direction means. If that appeals to you, dear reader, then you're certainly welcome to it. I have to say it doesn't do much for me.

It's only from within the folk mythology of progress, though, that we have no choice but to accept the endless prolongation of current trends. Right now, as individuals, we can choose to shrug and walk away from the latest hypermodern trinkets, and do something else instead.

2

THE DELUSION OF CONTROL

There's more involved in letting go of blind faith in progress, though, than setting aside any remaining enthusiasm one might happen to have for the next round of heavily marketed technological gimmickry. In particular, it's necessary to come to terms with the impact that the cult of progress has had on our ability to think. It's a truism of modern thought that progress has somehow made us smarter, but the evidence suggests otherwise.

It should have been obvious all along, for example, that treating the air as a gaseous sewer was a really bad idea, and in particular, that dumping billions upon billions of tons of heat-trapping gases into the atmosphere would change its capacity for heat retention in unwelcome ways. It should have been just as obvious that all the other ways we maltreat the only habitable planet we've got were guaranteed to end just as badly. That this wasn't obvious—that huge numbers of people find it impossible to realize that you can only wet your bed so many times before you have to sleep in a damp spot—deserves much more attention than it's received so far.

It really is a curious blindness, when you think about it. Since our distant ancestors climbed unsteadily down from the trees of late Pliocene Africa, the capacity to anticipate threats and do something about them has been central to the success of our species: a rustle in the grass might indicate the approach of a leopard; a series of unusually dry seasons might turn the local water hole into undrinkable mud. Those of our ancestors who paid attention to such things, and took constructive action in response to them, were more likely to survive and leave offspring than those who shrugged and went on with business as usual. That's why traditional societies around the world are hedged about with a dizzying assortment of taboos and customs meant to guard against every conceivable source of danger.

Somehow, though, we got from that to our present situation, where substantial majorities across the world's industrial nations seem unable to realize that something bad can actually happen to them, where thoughtstoppers of the "I'm sure they'll think of something" variety have, by and large, taken the place of serious thinking about the future, and where, when something bad does happen to someone, the immediate response is to find some way to blame the victim for what happened, so that everyone else can continue to believe that the same thing can't happen to them. It's not going too far to suggest that modern industrial society has become dangerously detached from the most basic requirements of collective survival.

That detachment shows most clearly, perhaps, in the way that certain projects have remained stuck in place as the next step in the supposedly unstoppable march of progress, while evidence piles up—and up and up—showing that they will never be realistic options. The most obvious example of this just now is nuclear fusion power.

Since the 1950s, a great many nuclear physicists have kept themselves employed by proclaiming commercial nuclear fusion power plants as the wave of the future. In just another twenty years, we've been told over and over again, clean, safe nuclear fusion plants will be churning out endless supplies of cheap electricity, if only the subsidies keep pouring in. After sixty years of unbroken failure, even politicians are starting to have second thoughts, but the fusion-power industry keeps at it. As respected science writer Charles Seife pointed out trenchantly in his excellent book *Sun in a Bottle: The Strange History of Fusion and the Science of Wishful Thinking*,[1] all this has more in common with the quest for perpetual motion than its overeager fans like to think.

Every few years the media carries yet another enthusiastic announcement that some new breakthrough has happened in the quest for fusion power. Of course, none of these widely ballyhooed breakthroughs ever amounts to a working fusion reactor capable of putting power into the grid, but let's let that pass for now, because the point I want to make is a different one. It's crucial to remember that whether fusion power is technically feasible is not the only question that matters. Another question, at least as important, is whether it's economically viable. That's not a question anyone in the fusion research industry wants to discuss, and there are good reasons for that.

The ITER project in Europe offers a glimpse at the answer.[2] ITER is the most complex, and thus also the most expensive, machine ever built by human beings—the latest estimate of the total cost has recently gone up from $14 billion to $17 billion, and if past performance is anything to go by, it will have gone up a good deal more before the scheduled completion in 2020. That stratospheric price tag results from the simple fact that six decades of hard work by physicists around the world, exploring scores of different approaches to fusion, have shown that any less expensive approach won't produce a sustained fusion reaction.

Take a moment to think through the economic consequences of that latter fact. Even if ITER manages a sustained fusion reaction, thousands of fusion power plants will have to be built if fusion is to make any difference at all. While commercial fusion reactors would doubtless cost less than ITER, they won't cost enough less to make fusion power economically viable; or to make the same point in other words, they won't produce electricity at a price that anyone can afford. Even if ITER succeeds in creating its "sun in a bottle," in other words, that fact will be an expensive laboratory curiosity, not a solution to the world's energy needs.

That so few people seem to be able to notice this, even as the bills for ITER rise and rise, shows just how deeply distorted our thinking has become under the influence of blind faith in progress. It's not merely unmentionable, but literally unthinkable, that something that's been defined by the media as the next step in the onward march of progress might be too expensive to afford. The unthinkable, however, is increasingly also the inescapable.

Believing in the Energy Fairy

Nor is nuclear fusion the only supposed wave of the future that's remained stuck permanently offshore, refusing to come rolling in to fulfill the fantasies loaded onto it by believers in progress. The assortment of technologies that are supposed to keep today's electrical grids and energy-intensive lifestyles powered by renewable means, as our remaining fossil fuel supplies deplete from under us, provide another painfully clear example. This can best be seen at work by considering the logic that underlies current alternative energy technologies.

Broadly speaking, there are two groups of people who talk about renewable energy these days. The first group consists of those people who believe that sun and wind can replace fossil fuels and enable modern industrial society to keep itself powered into the far future, using roughly the same amount of energy per capita it does today. The second group consists of people who actually live with renewable energy on a daily basis. It's been my repeated experience for years now that people belong to one of these groups or the other, but not to both.

As a general rule, in fact, the less direct experience a given person has living with solar and wind power, the more likely that person is to buy into a green cornucopianism that insists that sun, wind, and other renewable resources can provide everyone on the planet with a middle-class American lifestyle. Conversely, those people who have direct knowledge of the strengths and limitations of renewable energy—those, for example, who live in homes powered by sunlight and wind, without a fossil fuel-powered grid to cover up the intermittency problems—generally have no time for the claims of green cornucopianism and are the first to point out that relying on renewable energy means giving up a great many extravagant habits that most people in today's industrial societies consider normal.

Believers in a future powered by renewable energy, though, have not been willing to listen. Climate activist Naomi Oreskes, to cite one example out of many, published an article not long ago insisting that questioning whether renewable energy sources can power industrial society amounts to "a new form of climate denialism."[3] The same sort of angry rhetoric has begun to percolate all through the green end of our collective conversation: a shrill insistence that renewable energy sources are by definition able to fill in for fossil fuels, that of course the necessary buildout can be accomplished fast enough and on a large enough scale to matter, and that no one ought to be allowed to question these articles of faith.

The most instructive thing about this faith-based approach to energy is that it's not new. When I published my first book on peak oil back in 2007, the energy resource that was sure to save industrial civilization from itself was biofuels. Those of my readers who were paying attention to peak oil in those days will remember the grandiose and constantly reiterated pronouncements about the oceans of ethanol from American corn and the torrents of biodiesel from algae that were going to replace fossil fuels with all the cheap, abundant, carbon-neutral liquid fuel anyone could want. Those who raised annoying questions got reactions that swung across a narrow spectrum from patronizing putdowns to furious denunciation.

As it turned out, of course, the critics were right and the people who insisted that biofuels were the wave of the future were dead wrong. There were at least two problems, and both of them could have been identified—and in fact were identified—well in advance, by that minority who were willing to take a close look at the underlying data.

The first problem was that the numbers simply didn't work out. It so happens, for example, that if you grow corn using standard American agricultural methods and convert that corn into ethanol, using state-of-the-art industrial fermenters and the like, the amount of energy needed by that process is more than you get by burning the resulting ethanol. Thus corn ethanol is an energy sink, not an energy source. There are also crucial issues of scale; if you were to put every square inch of arable farmland in the world into biofuel crops, leaving none for such trivial uses as feeding the seven billion human beings on this planet, you still wouldn't get enough biofuel to replace the world's annual consumption of transportation fuels. Neither of these points were hard to figure out, and the second one was already well known in the 1970s,[4] but somehow the proponents of ethanol and biodiesel missed them both.

The second problem was a little more complex, but not enough so to make it impossible to figure out in advance. This was that biofuel production and consumption had consequences, and these were not necessarily as positive as they looked at first glance. Divert a significant fraction of the world's food supply into the fuel tanks of people in a handful of rich countries—and, of course, this is what all that rhetoric about fueling the world meant in practice—and the resulting spikes in food prices had disastrous impacts across the world, triggering riots in quite a number of countries and outright revolutions in more than one.

Meanwhile the global biosphere suffered as well, as rain forests in southeast Asia got clearcut so that palm oil plantations could supply the upper middle classes of Europe and America with supposedly sustainable biodiesel. It could have gotten much worse, but the underlying economics were so bad that within a few years companies started going broke at such a rate that banks stopped lending money for biofuel projects. Some of the most highly ballyhooed algal biodiesel projects turned out to be, in effect, pond scum Ponzi schemes;[5] and except for those enterprises that managed to get government subsidies, the biofuel boom went away.

It was promptly replaced by another energy resource that was sure to save industrial civilization: the hydrofracturing of oil- and gas-bearing shales. For quite a while, it was hard to find an energy-related website that wasn't full of grandiose diatribes glorifying hydrofracturing as a revolutionary new technology that, once it was applied to vast, newly discovered shale fields all over North America, was going to usher in a new era of US energy independence. Remember the phrase "Saudi America"? I certainly do.

Here again, there were two little problems with these claims, and the first was that once again the numbers didn't work out. Fracking wasn't a new technological breakthrough—it's been used on oil fields since the 1940s—and the "newly discovered" oil fields in North Dakota and elsewhere were nothing of the kind. They were found decades ago, and the amount of hydrocarbons they held did not justify the wildly overinflated claims made for them. Meanwhile the fracking industry, like the biodiesel industry, had impacts of its own that weren't limited to the torrents of new energy it was supposed to provide. All across the more heavily fracked parts of the United States, homeowners discovered that their tap water was so full of methane that they could ignite it with a match, while some had to deal with the rather more troubling consequences of earthquake swarms and miles-long trains of fracked fuels rolling across America's poorly maintained railroad network.

Things might have gotten much worse except, here again, the underlying economics of fracking were so bad that after a few years companies started going broke. Unless the industry figures out how to get government subsidies, fracking will shortly turn back into what it was before the current boom: one of several humdrum technologies used to scrape a little extra oil out from depleted oil fields. That, in turn, leaves the field clear for the next overblown "energy revolution"—and the focus of this upcoming round of energy hype will most likely be schemes meant to power the electrical grid with sun and wind.

In a way, that's convenient, because we don't have to wonder whether solar and wind power can evade the two problems that felled biofuels and fracking. That's already been settled; the research was done quite a while ago, and the answer is no.

To begin with, the numbers are just as problematic for solar and wind power as they were for biofuels and fracking. A thorough study of Spain's much-lauded solar energy program by Pedro Prieto and Charles A.S. Hall, for example, has worked out the net energy of large-scale solar photovoltaic systems on the basis of real-world data.[6] It's not pleasant reading if you happen to believe that today's lifestyles can be supported on sunlight. Prieto and Hall calculate that the energy return on energy invested (EROEI) of Spain's solar energy sector works out to 2.48—about a third of the figure suggested by less comprehensive estimates. That stunningly low figure may well explain why photovoltaic plants have consistently proven to be uneconomical unless they're propped up by government subsidies.

Similar challenges face every other attempt to turn renewable energy into a replacement for fossil fuels. Consider, for example, the study that showed on solid thermodynamic grounds that the total energy that can be taken from the planet's winds is a small fraction of what wind power advocates think they can get.[7] The logic here is irrefutable: there's a finite amount of energy in wind, and what you extract in one place won't turn the blades of another wind turbine somewhere else. Thus there's a hard upper limit to how much energy wind power can put into the grid—and it's not enough to provide more than a small fraction of the power needed by an industrial civilization.

Equally, renewables are by no means as environmentally benign as their more enthusiastic promoters claim. It's true that they don't dump as much carbon dioxide into the atmosphere as burning fossil fuels does—and it's indicative of the desperation of our times that talk about the very broad range of environmental blowbacks from modern industrial technologies has been supplanted by a much narrower focus on greenhouse gas–induced anthropogenic global warming, as though this is the only issue that matters. The technologies that turn sun and wind into grid electricity, however, involve very large volumes of rare metals, solvents, plastics, and other industrial products that have substantial carbon footprints of their own.

There are other problems of the same kind, some of which are already painfully clear. Many of those rare metals are affordable only because they're sourced from open-pit mines in the Third World worked by slave labor; the manufacture of most solvents and plastics involves the generation of a great deal of toxic waste, most of which inevitably finds its way into the biosphere; wind turbines are already racking up an impressive death toll among birds and bats—well, I could go on. Nearly all of modern industrial society's complex technologies are ecocidal to one fairly significant degree or another, and the fact that a few of them extract energy from sunlight or wind doesn't keep them from having a galaxy of nasty indirect environmental costs.

Point such details out to people in the contemporary green movement, and you can count on fielding an angry insistence that there's got to be some way to run industrial civilization on renewables, since we can't just keep on burning fossil fuels. I'm not at all sure how many of the people who make this sort of statement realize just how odd it is. It's as though they think some good fairy promised them that there would always be enough energy to support their current lifestyles, and the only challenge is figuring out where she hid it.

History's Actors—Or Not

Yet this straightforward blindness to the possibility that faith in progress may not be well founded is far from the only distortion that the cult of progress for its own sake has imposed on our thinking. Another, equally pervasive and at least as destructive, is the way that so many of us have come to treat the rest of the universe as though it was a machine. There's no shortage of examples here, but US foreign policy offers a particularly clear glimpse into the failure of reason in the age of progress.

As Paul R. Pillar pointed out in a thoughtful article,[8] the United States has a uniquely counterproductive notion of how negotiation works. Elsewhere on the planet, people understand that when you negotiate you're seeking a compromise in which you get whatever you most need out of the situation while the other side gets enough of its own agenda met to be willing to cooperate. To the US, by contrast, negotiation means that the other side complies with US demands, and that's the end of it. The idea that other countries might have their own interests, and might expect to receive some substantive benefit in exchange for cooperation with the US, has apparently never entered the heads of official Washington—and the absence of that idea has resulted in the cascading failures of US foreign policy in recent years.

The keynote of American foreign policy in the post-Soviet era, to cite only the most egregious of those failures, should have been driving as deep a wedge as possible between Russia and China, the only two nations large and strong enough to threaten the US–centric world order. Instead, in a display of political stupidity unmatched in modern times, the US government has gone out of its way to drive Russia and China into each other's arms, and recently accomplished the even more impressive feat of convincing the Islamic Republic of Iran to join the Russo-Chinese alliance. The result has been to turn a crippled Russia, a potentially friendly China, and an internationally isolated Iran into a Eurasian power bloc that rivals the US in its potential for global hegemony and is spoiling for a fight.

Let's step back from specifics, though, and notice the thinking that underlies the dysfunctional behavior in Washington. The people who think they're in charge inside the DC Beltway have lost track of the fact that Russia, China, and Iran have needs, concerns, and interests of their own, and aren't simply dolls that the US can pose at will. These other nations can, perhaps, be bullied by threats over the short term, but that's a strategy with a short shelf life. Successful diplomacy depends on giving the other guy reasons to want to cooperate with you, whereas demanding cooperation at gunpoint guarantees that the other guy is going to look for ways to shoot back.

The same sort of thinking in a different context underlies the brutal stupidity of American drone attacks in the Middle East. Some wag in the media pointed out a while back that the US went to war against an enemy 5,000 strong, we've killed 10,000 of them, and now there are only 20,000 left. That's a fair summary of the situation; the US drone campaign has been a total failure, having worked out consistently to the benefit of the Muslim extremist groups against which it's aimed, and yet nobody in official Washington seems capable of noticing this fact.

It's hard to miss the conclusion that the Bush and Obama administrations thought that in pursuing the drone-strike program, they were playing some kind of video game, which the United States can win if it can just rack up enough points. Notice the way that every report that a drone has taken out some al-Qaeda leader gets hailed in the media: hey, we nailed a commander, doesn't that boost our score by five hundred? In the real world, meanwhile, the indiscriminate slaughter of civilians by US drone strikes has become a core factor convincing Muslims around the world that the United States is just as evil as the jihadis claim, and thus sending young men by the thousands to join the jihadi ranks. Has anyone in Washington DC caught on to this straightforward arithmetic of failure? Surely you jest.

For that matter, I wonder how many of my readers recall the much-ballyhooed "surge" in Afghanistan some years back, during Barack Obama's first term. The "surge" was discussed at great length in the US media before it was enacted on Afghan soil; talking heads of every persuasion babbled learnedly about how many troops would be sent, how long they'd stay, and so on. It apparently never occurred to anybody in the Pentagon or the White House that the Taliban could visit websites, or just read newspapers, and get a pretty good idea of what the US forces in Afghanistan were about to do. That's exactly what happened, too; the Taliban simply hunkered down for the duration, and popped back up the moment the extra troops went home.

These examples of failure are driven by the same weird mental blindness: an inability to recognize that the other side has its own agenda and will respond to US actions in ways that further that agenda, rather than filling whatever role the US government chooses to assign it. It's the same failure of reasoning that leads so many people to assume that the biosphere is somehow obliged to give us all the resources we want and take all the abuse we choose to dump on it, without ever responding in ways that might inconvenience us.

We can sum up all these forms of acquired stupidity in a single sentence: a great many people these days seem to have lost the ability to grasp that the other side can learn.

The entire concept of learning has been so poisoned by certain other bad habits of contemporary thought that it's probably necessary to pause here. Learning, in particular, isn't the same thing as rote imitation. If you memorize a set of phrases in a foreign language, for example, that doesn't mean you've learned that language. To learn the language means to grasp the underlying structure so that you can come up with your own phrases and say whatever you want, not just what you've been taught to say.

To learn is to grasp the underlying structure of a given subject of knowledge so that the learner can come up with original responses to it. That's what Russia and China did; they grasped the underlying structure of US diplomacy, figured out that they had nothing to gain by cooperating with that structure, and came up with a creative response, which was to ally against the United States. That's also what the jihadis and the Taliban are doing in the face of US military activity. If life hands you lemons, as the saying goes, make lemonade; if the US hands you drone strikes that routinely slaughter noncombatants, you can make very successful propaganda out of it—and if the US hands you a surge, you roll your eyes, hole up in your mountain fastnesses, and wait for the Americans to get bored or distracted, knowing that this won't take long. That's how learning works, but that's something that US planners seem congenitally unable to take into account.

The same analysis, interestingly enough, makes just as much sense when applied to nonhuman nature. As Ervin László pointed out a long time ago in his *Introduction to Systems Philosophy*,[9] any sufficiently complex system behaves in ways that approximate intelligence. Consider the way that bacteria respond to antibiotics. Individually, bacteria have no intelligence worth measuring, but their behavior on the species level shows an eerie similarity to learning; faced with antibiotics, a species of bacteria "tries out" different biochemical approaches until it finds one that sidesteps the antibiotic. In the same way, insects and weeds "try out" different responses to pesticides and herbicides until they find whatever allows them to munch on crops or flourish in the fields no matter how much poison the farmer sprays on them.

We can even apply the same logic to the environmental crisis as a whole. Complex systems tend to seek equilibrium and will respond to anything that pushes them away from equilibrium by pushing back the other way. Any field biologist can show you plenty of examples: if conditions allow more rabbits to be born in a season, for instance, the population of hawks and foxes rises accordingly, reducing the rabbit surplus to a level the ecosystem can support. As humanity has put increasing pressure on the biosphere, the biosphere has begun to push back with increasing force, in an increasing number of ways; is it too much to think of this as a kind of learning in which the biosphere "tries out" different ways to balance out the abusive behavior of humanity and will eventually find one that works?

It's a commonplace of modern thought that natural systems can't learn. The notion that nature is static, timeless, and unresponsive, a passive stage on which human beings alone play active roles, is welded into modern thought, unshaken even by the realities of biological evolution or the rising tide of evidence that natural systems are in fact quite able to adapt their way around human meddling. There's a long and complex history to the notion of passive nature, but that's a subject for another day. What interests me just now is that since 1990 or so, the governing classes of the United States, and some other Western nations as well, have applied the same, frankly delusional, logic to everything in the world other than themselves.

"We're an empire now, and when we act, we create our own reality," neoconservative guru Karl Rove is credited as saying to reporter Ron Suskind. "We're history's actors, and you, all of you, will be left to just study what we do."[10] That's thinking that governs the US government these days. The president says we're in a recovery, and if the economy fails to act accordingly, why, industrious flacks in government and the media churn out elaborately fudged statistics to erase that unwelcome reality. That history's self-proclaimed actors might turn out to be just one more set of flotsam awash on history's vast tides has never entered their darkest dream.

Let's step back from specifics again, though. What's the source of this frankly bizarre belief system—the delusion that leads politicians to think that they create reality, and that everyone and everything else can only fill the roles they've been assigned by history's actors? My proposal is that it comes from a simple but remarkably powerful fact, which is that the people in question, along with most people in the privileged classes of the industrial world, spend most of their time dealing with machines rather than living things.

We can define a machine as a subset of the universe that's been deprived of the capacity to learn. The whole point of building a machine is that it does what you want, when you want it, and nothing else. Flip the switch on, and it turns on and goes through whatever rigidly defined set of behaviors it's been designed to do. Flip the switch off, and it stops. The machine may be fitted with controls so you can manipulate its behavior in various ways; nowadays, especially, the set of behaviors assigned to it may be extremely complex. There's no inner life behind the facade, though. It can't learn, and to the extent that it pretends to learn, all that comes out of it is the product of the sort of rote memorization described above as the antithesis of learning.

A machine that learned would be capable of making its own decisions and coming up with a creative response to your actions. That's the opposite of what machines are meant to do, because that response might well involve frustrating your intentions so that the machine can get what it wants instead. That's why the trope of machines going to war against human beings has so large a presence in popular culture: it's exactly because we expect machines not to act like people, not to pursue their own needs and interests, that the thought of machines acting the way we do gets so reliable a frisson of horror.

The delusion of control—the conviction, apparently immune to correction by mere facts, that some human beings are "history's actors" and the rest of the cosmos is obliged to respond in a mechanical fashion to their actions—is a pervasive mental health problem in the modern world. It goes unrecognized because it's so widespread, because the popular ideologies of the present day support and encourage it, and above all because an environment too lavishly stocked with machines fosters the habits of thought that make the delusion of control seem to make sense.

The Prosthetic Society

Very often these days, as noted earlier, things labeled "more advanced," "more progressive," and the like are very often less satisfactory and less effective at meeting human needs than the allegedly more primitive technologies they replaced. By and large, in fact, today's technologies fail systematically at meeting certain human needs. One of the central reasons for that failure is that the peak of technological complexity in our time can also be described accurately enough as peak meaninglessness.

I'd like to take the time to unpack that phrase. In the most general sense, technologies can be divided into two broad classes, which we can respectively call tools and prosthetics. The difference is a matter of function. A tool expands human potential, giving people the ability to do things they couldn't otherwise do. A prosthetic, on the other hand, replaces human potential, doing something that under normal circumstances, people can do just as well for themselves. Most discussions of technology these days focus on tools, but the vast majority of technologies that shape the lives of people in a modern industrial society are not tools but prosthetics.

Prosthetics have a definite value, to be sure. Consider an artificial limb, the sort of thing on which the concept of technology-as-prosthetic is modeled. If you've lost a leg in an accident, say, an artificial leg is well worth having; it replaces a part of ordinary human potential that you don't happen to have any more, and it enables you to do things that other people can do with their own leg. Imagine, though, that some clever marketer were to convince people to have their legs cut off so that they could be fitted for artificial legs. Imagine, furthermore, that the advertising for artificial legs became so pervasive, and so successful, that nearly everybody became convinced that human legs were hopelessly old-fashioned and ugly, and rushed out to get their legs amputated so they could walk around on artificial legs.

Then, of course, the manufacturers of artificial arms got into the same sort of marketing, followed by the makers of sex toys. Before long you'd have a society in which most people were gelded quadruple amputees fitted with artificial limbs and rubber genitals, who spent all their time talking about the wonderful things they could do with their prostheses. Only in the darkest hours of the night, when the TV was turned off, might some of them wonder why it was that a certain hard-to-define numbness had crept into all their interactions with other people and the rest of the world.

In a very real sense, that's the way modern industrial society has reshaped and deformed human life. Take any human activity, however humble or profound, and some clever marketer has found a way to insert a piece of technology between the person and the activity. You can't simply bake bread—a simple, pleasant activity that people have done themselves for thousands of years using their hands and a few simple handmade tools. No, you have to have a bread machine, into which you dump a prepackaged mix and some liquid, push a button, and stand there being bored while it does the work for you, if you don't farm out the task entirely to a bakery and get the half-stale industrially extruded product that usually passes for bread these days.

Of course, the bread machine manufacturers and the bakeries pitch their products to the clueless masses by insisting that nobody has time to bake their own bread any more. A long time ago, in his book *Energy and Equity*, Ivan Illich pointed out the logical fallacy central to such claims: using a bread machine or buying from a bakery is faster only if you don't count the time you have to spend earning the money needed to pay for it, power it, provide it with overpriced prepackaged mixes, repair it, clean it, and so on.[11] Illich's discussion focused on automobiles; he pointed out that if you take the distance traveled by the average American auto in a year, and divide that by the total amount of time spent earning the money to pay for the auto, fuel, maintenance, insurance, et cetera, plus all the other time eaten up by tending to the auto in various ways, the average American car goes about 3.5 miles an hour: about the same pace, that is, that an ordinary human being can walk.

The claim that technology saves time and labor seems to make sense, in other words, only if you ignore a whole series of externalities—in this case, the time you have to put into earning the money to pay for the technology and into coping with whatever requirements, maintenance needs, and side effects the technology has. Have you noticed that the more "time-saving technologies" you bring into your life, the less free time you have? This is why—and it's also why the average medieval peasant worked shorter hours, had more days off, and kept a larger fraction of the value of his labor than you do.[12]

Something else is being externalized by prosthetic technology, though. What are you doing, really, when you use a bread machine? You're not baking bread; the machine is doing that. You're dumping a prepackaged mix and some water into a machine, closing the lid, pushing a button, and going away to do something else. Fair enough—but what is this "something else" that you're doing? In today's industrial societies, odds are you're going to go use another piece of prosthetic technology, which means that once again, you're not actually doing anything. A machine is doing something for you. You can push that button and walk away, but again, what are you going to do with your time? Use another machine?

The machines that industrial society use to give this infinite regress somewhere to stop—televisions, video games, and computers hooked up to the internet—simply take the same process to its ultimate extreme. Whatever you think you're doing when you're sitting in front of one of these things, what you're actually doing is staring at little colored pictures on a glass screen and maybe pushing some buttons from time to time. All things considered, this is a profoundly boring activity, perhaps the most boring activity human beings have ever pursued, which is why the little colored pictures jump around all the time. That's done to keep your nervous system so far off balance that you don't notice just how tedious it is to spend hours at a time staring at little colored pictures on a screen.

I can't help but laugh when people insist that the internet is an information-rich environment. Quite the contrary, all you get from it is the very narrow trickle of verbal, visual, and auditory information that can squeeze through the digital bottleneck and turn into little colored pictures on a glass screen. The best way to experience this is to engage in a media fast—a period in which you deliberately cut yourself off from all electronic media for a week or more, preferably in a quiet natural environment. If you do that, you'll find that it can take two or three days, or even more, before your numbed and dazzled nervous system recovers far enough that you can begin to tap in to the ocean of sensory information and sensual delight that surrounds you at every moment. It's only then, furthermore, that you can start to think your own thoughts and dream your own dreams, instead of just rehashing whatever the little colored pictures tell you.

A movement of radical French philosophers back in the 1960s, the Situationists, argued that modern industrial society is basically a scheme to convince people to hand over their own human capabilities to the industrial machine so that imitations of those capabilities can be sold back to them at premium prices.[13] It was a useful analysis then, and it's even more useful now, when the gap between realities and representations has become even more drastic than it was in the 1960s. These days, as often as not, what gets sold to people isn't even an imitation of some human capability but an abstract representation of it, an arbitrary marker with only the most symbolic connection to what it represents.

This is one of the reasons why I think it's deeply mistaken to claim that Americans are materialistic. Americans are arguably the least materialistic people in the world. No actual materialist—no one who had the least appreciation for actual physical matter and its sensory and sensuous qualities—could stand the vile plastic tackiness of America's built environment and consumer economy for a fraction of a second. Americans don't care in the least about matter; they're happy to buy even the most ugly, uncomfortable, shoddily made, and absurdly overpriced consumer products you care to imagine, so long as they've been convinced that having those products symbolizes some abstract quality they want, such as happiness, freedom, sexual pleasure, or what have you.

Then they wonder in the darkest hours of the night, when the TV is turned off, why all the things that are supposed to make them happy and satisfied somehow never manage to do anything of the kind. Of course, there's a reason for that, too, which is that happy and satisfied people don't keep on frantically buying products in a quest for happiness and satisfaction. Still, the little colored pictures keep showing them images of people who are happy and satisfied because they guzzle the right brand of tasteless fizzy sugar water, pay for the right brand of shoddily made half-disposable clothing, and keep watching the little colored pictures: that last above all else. "Tune in tomorrow" is the most important product that every media outlet sells, and they push it every minute of every day on every stop and key.

That is to say, between my fantasy of voluntary amputees eagerly handing over the cash for the latest models of prosthetic limbs, and the reality of life in a modern industrial society, the difference is simply in the less permanent nature of the alterations imposed on people here and now. It's easier to talk people into amputating their imaginations than it is to convince them to amputate their limbs, but it's also a good deal easier to reverse the surgery.

What gives this even more importance than it would otherwise have, in turn, is that all this is happening in a society that's hopelessly out of touch with the realities that support its existence and that relies on bookkeeping tricks of the sort discussed earlier to maintain the fantasy that it's headed somewhere other than history's well-used compost bin. The externalization of the mind and the imagination plays just as important a role in maintaining that fantasy as the externalization of costs—and the cold mechanical heart of the externalization of the mind and imagination is *mediation*, the insertion of technological prosthetics into the space between the individual and the world.

The usual rhetoric that surrounds this process of mediation claims that it's the inevitable result of progress and the onward march of technology. That's profoundly deceptive, and a case could be made that it's deliberately so, but like most deceptive rhetoric it expresses important truths in a backhanded manner. One truth that it expresses is that, by and large, the changes in technology our society calls "progress" have increased the layers of mediation separating the individual from the world and have driven us all further toward the peak meaninglessness of today.

The points made in this chapter have a further implication. The self-defeating habits of thought I've sketched out here aren't "primitive" or "backward." They aren't things that past generations had to deal with. We didn't get them by failing to move forward fast enough. Quite the contrary, we *progressed* into them. The failure to grasp that progress is not inevitable and that other people and things can learn, the obsessive use of the machine as a metaphor, the descent into a maze of abstractions created and projected by modern media—these are the brand-new, innovative, up-to-date problems that we've made for ourselves.

The conventional wisdom of our time insists that the only possible solution to these problems is even more progress. As we'll see, though, there's a more promising option.

3

GOING FORWARD BY GOING BACK

THE LANDSCAPE SKETCHED out in the first two chapters of this book is a familiar one to most people nowadays, whether or not they've taken the time to put the details into context. Even those who profess an unshaken faith in progress have to cope with the new and improved products that are new but not improved, the upgrades that might better be described as downgrades, and the rest of it. Despite the enforced cheerfulness of the media, for that matter, most people in today's industrial nations realize at some level that things are going very, very wrong—that policies that were supposed to provide peace and prosperity are reliably generating the opposite, and the problems that were supposed to be solved by those policies have gotten steadily worse. The question that remains is what can be done about it all.

An astonishing number of people these days seem unable to imagine the possibility that the universe might ignore our delusion of control and do things to us that we don't want. The problem we face now, as already noted, is precisely that the unimaginable is now our reality. For just that little bit too long, too many people have insisted that we didn't need to worry about the absurdity of pursuing limitless growth on a finite and fragile planet, and now the blowback from all those externalized costs discussed in Chapter One is beginning to pick up momentum.

This, too, is a consequence of the problematic thinking that's resulted from faith in progress. Too many people have repeated the mantra "They'll think of something" or assumed that chattering on internet forums about this or that or the other piece of technological vaporware was doing something concrete about the future. For just that little bit too long, again, not enough people were willing to do anything that mattered, and now impersonal factors have climbed into the driver's seat, having mugged all seven billion of us and shoved us into the trunk.

That puts hard limits on what can be done in the short term. This doesn't mean the world is about to end. It means that for the foreseeable future, most of us will be busy coping with the multiple impacts of assorted political and economic crises on our lives and those of our families, friends, communities, and employers at a time when political systems over much of the industrial world have frozen up into gridlock, the simmering wars in the Middle East and much of the Third World seem more than usually likely to boil over, and the twilight of the Pax Americana is destabilizing the international order. Exactly how that's going to play out is anyone's guess, but no matter what happens, it's unlikely to be pretty.

One of the gifts of crisis, though, is the way it changes the boundaries of what's possible and thinkable. Read Barbara Tuchman's brilliant *The Proud Tower* and you'll see how many of the unquestioned certainties of 1914 were rotting in history's compost bucket by the time 1945 rolled around, and how many ideas that had been on the outermost fringes before the First World War that had become plain common sense after the Second. It's a common phenomenon, and I suggest it's time to get ahead of the curve by proposing something that's utterly unthinkable today but may well be thinkable, and even necessary, in the not too distant future.

What do I have in mind? Deliberate technological regression as a matter of public policy.

Imagine, for a moment, that an industrial nation were to downshift its technological infrastructure to roughly what it was in 1950. That would involve a drastic decrease in energy consumption per capita, both directly—people used a lot less energy of all kinds in 1950—and indirectly—goods and services took much less energy to produce then, too. It would involve equally sharp decreases in the per capita consumption of most resources. It would also involve a sharp *increase* in jobs for the working classes—a great many things currently done by computers and industrial robots were done by human beings in those days, and so there were a great many more paychecks going out of a Friday to pay for the goods and services that ordinary consumers buy. Since a steady flow of paychecks to the working classes is one of the major things that keep an economy stable and thriving, this has certain obvious advantages, but we can leave those alone for now.

Now, of course, the change just proposed would involve certain changes from the way we do things. Air travel in the 1950s was extremely expensive—the well-to-do in those days were called "the jet set," because that's who could afford tickets—and so everyone else had to put up with fast, reliable, energy-efficient railroads to get from place to place. Computers were rare and expensive, which meant once again that more people got hired to do jobs, and also meant that when you called a utility or a business, your chance of getting a human being who could help you with whatever problem you might have was considerably higher than it is today.

Lacking the internet, people had to make do instead with their choice of scores of AM and shortwave radio stations, thousands of general and special-interest newspapers and magazines, and local libraries bursting at the seams with books—in America, at least, the 1950s was the golden age of the public library, and many small towns had collections better than the ones you find in big city libraries these days. Meanwhile, the folks who like looking at pictures of people with their clothes off, and who play a huge and usually unmentioned role in paying for the internet today, had to settle for monthly magazines, mail-order houses that shipped their products in plain brown wrappers, and tacky stores in the wrong end of town. (For what it's worth, this didn't seem to inconvenience them any.)

Thinking the Unthinkable

I'm quite aware that such a downshift is unthinkable, and we'll get to the superstitious horror that lies behind that reaction in a bit. First, though, let's talk about the obvious objections. Would it be possible? Of course. Much of it could be done by simple changes in the tax code. Right now, in the United States, a galaxy of perverse tax incentives penalize employers for hiring people and reward them for replacing employees with machines. Any employer who hires someone, for example, must not only cover the costs of a wage or a salary but also pay into Social Security, unemployment insurance, the fund that compensates workers injured on the job, and so on. All these discourage employment by making it more expensive to hire people than it would otherwise be.

Installing machines to replace employees removes these costs, and the tax code imposes no comparable financial burden on automation. In fact, money spent on automation counts as a capital investment that can be depreciated over years to come, reducing the employer's tax burden. This acts as a subsidy for automation. Change the tax code so that spending money on wages, salaries, and benefits up to a certain comfortable threshold makes financial sense for employers, while automation is taxed to cover the costs of increasing unemployment, and that in itself will likely cover much of the necessary ground.

A revision in trade policy would do most of the rest of what's needed. What's jokingly called "free trade" benefits the rich at everyone else's expense and would best be replaced by sensible tariffs to support domestic production against the sort of predatory export-driven mercantilism that dominates the global economy these days. Add to that high tariffs on technology imports and strip any post–1950-level technology of the lavish subsidies that fatten the profit margins of the welfare-queen corporations in the *Fortune* 500, and you're basically there. Other policies might be needed to complete the process, but most of it could be done simply enough by stripping away the government incentives that encourage high technology and replacing them with incentives to use older and simpler machines, or to hire workers instead.

One of the great advantages of technological regression is that it involves far fewer unwelcome surprises and dead ends than trying to develop some new technological system. Figuring out what works, how to implement it, what will be needed and how much it will cost is simple. All that's needed is a glance back down memory lane, with the help of an abundance of surviving records. In the case of 1950 technology, there's also a fair amount of living memory to draw from. Thus it's wholly practicable, once the decision is made to do it.

Would there be downsides? Of course. Every technology and every set of policy options has its downsides. The rhetoric of progress claims, in effect, that it's unfair to take the downsides of new technologies or the corresponding upsides of old ones into consideration when deciding whether to replace an older technology with a newer one. An even more common bit of rhetoric claims that you're not supposed to decide. Once a new technology shows up, you're supposed to run bleating after it like everyone else, without asking any questions at all.

Current technology has immense downsides. Future technologies are going to have them, too—it's only in sales brochures and science fiction stories that any technology anywhere lacks problems. Thus the mere fact that 1950 technology has problematic features, too, is not a valid reason to dismiss technological retrogression. The question that needs to be asked, however unthinkable it might be, is whether, all things considered, it's wiser to accept the downsides of 1950 technology in order to have a working technological suite that can function on much smaller per capita inputs of energy and resources, and thus a much better chance to get through the age of limits ahead, than to have today's far more extravagant and brittle technological infrastructure.

To true believers in the religion of progress, though, this is unthinkable. To them, the past is the bubbling pit of eternal damnation from which the surrogate messiah of progress is perpetually saving us, and the future is the radiant heaven into whose portals the faithful hope to enter in good time. Nothing, but nothing, stirs up shuddering superstitious horror in the minds of the majority these days like the thought of "going back." Even if the technology of an earlier day is better suited to a future of energy and resource scarcity than the infrastructure we've got now, even if the technology of an earlier day actually does a better job of many things than what we've got today, "We can't go back!" is the anguished cry of the faithful. They've been so thoroughly bamboozled by the propagandists of progress that they never stop to think that, why, yes, they can, and there are valid reasons why they might even decide that it's the best option open to them.

There's a rich irony in the fact that alternative and avant-garde circles tend to be even more obsessively fixated on the dogma of linear progress than the supposedly more conformist masses. That's one of the sneakiest features of the myth of progress. When people get dissatisfied with the status quo, the myth convinces them that the only option they've got is to keep on doing exactly what everyone else is doing, but take it a little further than anyone else has gotten yet. What starts off as rebellion thus gets co-opted into perfect conformity, and society continues to march mindlessly along its current trajectory without ever asking the obvious questions about what might be waiting at the far end.

That's the thing to keep in mind about progress. All the word means, as we'll see, "continued movement in the same direction."

If the direction was a bad idea to start with, or if it's passed the point at which it still made sense, continuing to trudge blindly onward into the gathering dark may not be the best idea in the world. Break out of that mental straitjacket, and the range of possible futures broadens out immeasurably.

It may be, for example, that technological regression to the level of 1950 turns out to be impossible to maintain over the long term. If the technologies of 1920 can be supported on the modest energy supply we can count on getting from renewable sources, for example, something like a 1920 technological suite might be maintained over the long term, without further regression. It might turn out instead that a renewable equivalent of 1880s technology, using solar steam engines of the kind developed in the late nineteenth century,[1] might be the most complex technology that can be supported over the long term.

It might be the case, for that matter, that something like the technological infrastructure the United States had in 1820, with windmills and water wheels as the prime movers of industry, canal boats as the core domestic transport technology, and most of the population working on small family farms to support very modest towns and cities, is the fallback level that can be sustained indefinitely. Does that last option seem unbearably depressing? Compare it to an even more likely scenario—the consequences if the world's industrial societies gamble their survival on a great leap forward to some unproven energy source, which doesn't live up to its billing, and leaves seven billion people twisting in the wind without any working technological infrastructure at all—and you may find that it has its good points.

Back in 2005, the Hirsch report showed that any attempt to address the impending collision with the hard ecological limits of a finite planet had to get under way at least twenty years before the peak of global conventional petroleum reserves if there was to be any chance of avoiding massive disruptions.[2] In the wake of that announcement, instead of dealing with the hard realities of our predicament, the industrial world panicked and ran the other way, with the United States well in the lead. Strident claims that ethanol—er, solar energy—um, biodiesel—okay, wind—well, fracking, then—would provide enough cheap energy to replace the world's rapidly depleting reserves of oil, coal, and natural gas took the place of a serious energy policy, while conservation, the one thing that might have made a difference, was as welcome as garlic aioli at a convention of vampires.

That stunningly self-defeating response had a straightforward cause, which was that everyone except a few of us on the fringes treated the whole matter as though the issue was how to keep the great onward march of progress going in exactly the same direction as before. Since that question has no meaningful answer, questions that could have been answered—for example, How do we get through the impending mess with at least some of the achievements of the past three centuries intact?—never got asked at all. Thus years have been wasted trying to come up with answers to the wrong question, and most of the doors that were still open in 2005 have been slammed shut by events since that time.

Fortunately, there are still a few possibilities open even this late in the game. More fortunate still, the ones that will likely matter most don't require America's serenely clueless political class to do something useful for a change. They depend, rather, on personal action, beginning with individuals, families, and local communities and spiraling outward from there to shape the future on wider and wider scales.

Possibilities for Constructive Action

Two of those possibilities have already been discussed at length in my earlier books. The first can be summed up simply enough in a cheery sentence: "Collapse now, and avoid the rush!" In an age of economic contraction, nothing is as self-defeating as the attempt to prop up extravagant lifestyles that the real economy of goods and services will no longer support. Those who thrive in such times are those who downshift ahead of the economy, accept a much less extravagant personal lifestyle, take the resources freed up by this maneuver, and apply them to the costs of transition to less absurd ways of living.

The point of this project isn't limited to its advantages on the personal scale, though these are substantial. It's been demonstrated over and over again that personal example is more effective than verbal rhetoric at laying the groundwork for collective change. Many people stay penned in the increasingly unsatisfying and unproductive lifestyles sold to them by the media simply because they can't imagine a better alternative. Those people who collapse ahead of the rush, and demonstrate that it's entirely possible to have a humane and decent life on a small fraction of the usual American resource footprint, are already functioning as early adopters.

The second possibility is considerably more complex and resists summing up so neatly. In my book *Green Wizardry*, I sketched out the toolkit of concepts and approaches that were central to the appropriate technology movement back in the 1970s, when I had my original education in sustainability. I argued then, and still believe now, that by some combination of genius and sheer dumb luck the pioneers of that movement managed to stumble across a set of approaches to the work of sustainability that are much better suited to the needs of our time than anything that's been proposed since then.

Among the most important features of the "green wizardry" of appropriate tech is the fact that those who want to get working on it don't have to wait for anyone else to lift a finger. Government grants and the like aren't necessary, or even useful. From its roots in the '60s counterculture, the appropriate-tech scene inherited a focus on do-it-yourself projects that could be done with hand tools, hard work, and not much money. In an age of economic contraction, that makes even more sense than it did back in the day, and the ability to keep yourself and others warm, dry, fed, and provided with many of the other needs of life without potentially lethal dependencies on today's baroque technostructures has much to recommend it.

Crucially, too, appropriate tech is not limited to those who can afford a farm in the country. Many of the most ingenious and useful appropriate-tech projects were developed by and for people living in ordinary homes and apartments, with a small backyard or no soil at all available for gardening. The most important feature of appropriate tech, though, is that the core elements of its toolkit—intensive organic gardening and small-scale animal husbandry, home-scale solar thermal technologies, energy conservation, and the like—will still make sense long after fossil fuels have gone the way of the dinosaurs. Getting these techniques into as many hands as possible now is thus not just a matter of cushioning the impacts of the impending era of crisis; it's also a way to start building the sustainable world of the future right now.

The strategy of deliberate technological regression explored in this book is a third option, one that can be pursued alongside the two just named. It gains its power from a feature of progress that's not always grasped by its critics, much less by those who've turned faith in progress into the established religion of our time. Very few new technologies meet human needs that weren't already being met, and so the arrival of a new technology generally leads to the abandonment of an older technology that did the same thing. The difficulty here is that new technologies nowadays are inevitably more dependent on global technostructures, and the increasingly brittle and destructive economic systems that support them, than the technologies they replace.

This is the basis for the externality trap sketched out in Chapter One. This trap isn't a mere theoretical possibility. It's an everyday reality, especially but not only in North America today. Plenty of forces drive the rising spiral of economic, social, and environmental disruption that's shaking the industrial world right down to its foundations, but among the most important is precisely the unacknowledged impact of externalized costs on the whole systems that support the industrial economy.

It's fashionable these days to insist that increasing technological complexity and integration will somehow tame that rising spiral of crisis, but the externality trap suggests that exactly the opposite is the case—that the more complex and integrated technologies become, the more externalities they will generate, and the heavier a burden externalities will place on society, the economy, and the biosphere. Nor is increased complexity necessarily a good thing. It's precisely because technological complexity makes it easy to ignore diminishing returns and externalized costs that progress becomes its own nemesis.

I know that it's heresy to suggest that progress isn't infallibly beneficent, and heresy twice over to suggest that progress will inevitably terminate itself with extreme prejudice. I can't help that. It so happens that in most declining civilizations, ours included, the things that most need to be said are the things that, by and large, nobody wants to hear. That being the case, let's make it three for three and point out that the externality trap is a problem rather than a predicament. The difference is that problems can be solved, while predicaments can only be faced. We don't have to keep loading an ever-increasing burden of diminishing returns and externalized costs on the whole systems that support us—which is to say, we don't have to keep increasing the complexity and integration of our technologies. We can stop adding to the burden; we can even go the other direction.

Of course, suggesting that, even thinking it, is far more heretical than anything I've mentioned yet. I'm reminded here of a bit of technofluff that appeared in the Canadian media a while back,[3] claiming to present a radically pessimistic view of the next ten years. Of course, being a product of the mainstream media, it had as much in common with actual pessimism as lite beer has with a pint of good brown ale; the worst thing the author, Douglas Coupland, is apparently able to imagine is that industrial society will keep on doing what it's doing now—though the fact that more of what's happening now apparently counts as radical pessimism these days is an interesting point.

The detail of this particular Dystopia Lite that deserves attention here, though, is Coupland's dogmatic insistence that "you can never go backward to a lessened state of connectedness." That's a common bit of rhetoric out of the mouths of tech geeks these days, to be sure, but it isn't even remotely true. I know quite a few people who used to be active on social media and have dropped the habit. I know others who used to have high-end smartphones and went back to ordinary cell phones, or even to a plain land line, because they found that the costs of excess connectedness outweighed the benefits. Technological downshifting is already a rising trend, and there are very good reasons for that.

Most people find out at some point in adolescence that there really is such a thing as drinking too much beer. Many people are slowly realizing that the same thing is true of connectedness, and of the other prominent features of today's technological fashion statements. One of the data points that gives me confidence in that analysis is the way that people like Coupland angrily dismiss the possibility. Part of his display of *soi-disant* pessimism is the insistence that within a decade, people who don't adopt the latest technologies will be dismissed as passive-aggressive control freaks. Of course, that label could be turned the other way just as easily, but the point I want to make here is that nobody gets so bent out of shape about behaviors that are mere theoretical possibilities. Clearly, Coupland and his geek friends are already contending with people who aren't interested in conforming to the fashions of the technosphere.

It's not just geek technologies that are coming in for that kind of rejection, either. These days, in the town where I live, teenagers whose older siblings used to go hotdogging around in cars ten years ago are doing the same thing on bicycles today. Granted, I live in a down-at-heel old mill town in the north central Appalachians, but there's more to it than that. For a lot of these kids, the costs of owning a car outweigh the benefits so drastically that cars aren't cool any more. One consequence of that shift in cultural fashion is that these same kids aren't contributing anything like as much to the buildup of carbon dioxide in the atmosphere or to the other externalized costs generated by car ownership.

Deliberate technological regression is thus something that can be pursued constructively by individuals. Partly this is because the deathgrip of failed policies on the political and economic order of the industrial world is tight enough that any significant change these days has to start down here at the grassroots level, with individuals, families, and communities, if it's going to get anywhere at all. Partly, it's because technological regression, like anything else that flies in the face of the media stereotypes of our time, needs the support of personal example in order to get a foothold. Partly, it's because older technologies, being less vulnerable to the impacts of whole-system disruptions, will still be there meeting human needs when the grid goes down, the economy freezes up, or something really does break the internet, and many of them will still be viable when the fossil fuel age is a matter for the history books.

Still, there's another aspect, and it's one that the essay by Douglas Coupland mentioned above managed to hit squarely: the high-tech utopia ballyhooed by the first generation or so of internet junkies has turned out in practice to be a good deal less idyllic, and a good deal more dystopian, than its promoters claimed. All the wonderful things we were supposedly going to be able to do turned out in practice to consist of staring at little pictures on glass screens and pushing buttons, and as already noted, these activities just aren't that interesting. The people who are downshifting to a less connected lifestyle have noticed this.

What's more, a great many more people—the kids hotdogging on bikes here in Cumberland among them—are weighing the costs and benefits of complex technologies with cold eyes and deciding that an older, simpler technology less dependent on global technosystems is not just more practical but also, and importantly, more fun. True believers in the cyberfuture will doubtless object to that last point, but the deathgrip of failed ideas on societies in decline affects fashionable intellectuals most of all and makes them proclaim the imminent arrival of the future's rising waters when the tide's already turned and is flowing back out to sea.

Thus it's entirely possible that we could be heading toward a future in which people will roll their eyes when they think of Twitter, texting, 24/7 connectivity, and the rest of today's technofetishism—like, dude, all that stuff is *so* twenty-teens! Meanwhile, those who adopt the technologies and habits of earlier eras, whether that adoption is motivated by mere boredom with jerky colored pictures on little glass screens or by some more serious set of motives, may actually be on the cutting edge.

The Benefits of Negative Progress

In this context it's worth noting a common euphemism that's found its way into current chatter about the economy. Turn to any set of talking-head economists in the media discussing the downside of today's trends, and you're likely to encounter the phrase "negative growth." Negative growth? Why, yes, that's the opposite of growth.

Of course, the English language has no shortage of perfectly clear words for the opposite of growth. "Decline" comes to mind; so does "decrease," and so does "contraction." Would it have been so very hard for the talking heads to draw in a deep breath and actually come right out and say "The US economy has contracted" or "GDP has decreased" or even "We're currently in a state of economic decline"? Come on, economists, you can do it!

But of course they can't. Economists are supposed to provide, shall we say, negative clarity when discussing certain aspects of contemporary economic life, and those who become talking heads in the media are even more subject to this rule. Among the things about which they're supposed to be negatively clear, two are particularly relevant here; the first is that economic contraction happens, and the second is that that letting too much of the national wealth end up in too few hands is a very effective way to cause economic contraction. The logic here is uncomfortably straightforward—an economy that depends on consumer expenditures prospers only if consumers have money to spend—but talking about that equation would cast an unwelcome light on the culture of mindless kleptocracy entrenched these days at the upper end of the North American socioeconomic ladder. So we get to witness the mass production of negative clarity about one of the main causes of negative growth.

It's enticing to think of other uses for this convenient mode of putting things. I can readily see it finding a role in health care: "I'm sorry, ma'am," the doctor says, "but your husband is negatively alive"; in sports: "Well, Joe, unless the Orioles can cut down that negative lead of theirs, they're likely headed for a negative win"; and in the news: "The situation in Syria is shaping up to be yet another negative triumph for US foreign policy." For that matter, it's time to update one of the more useful proverbs of recent years: what do you call an economist who makes a prediction? Negatively right.

Come to think of it, we might as well borrow the same turn of phrase for the deliberate adoption of older, simpler, more independent technologies in place of today's newer, more complex, and more interconnected ones. I've been talking about that project so far under the negatively mealy-mouthed label "intentional technological regress," but hey, why not be cool and adopt the latest fashion? The same thing could as well be called "negative progress." Since negative growth sounds like just another kind of growth, negative progress ought to pass for another kind of progress, right?

With this in mind, I'd like to talk about some of the reasons that individuals, families, organizations, and communities, as they wend their way through today's cafeteria of technological choices, might want to consider loading up their plates with a good hearty helping of negative progress.

Let's start by returning to one of the central points raised earlier, the relationship between progress and the production of externalities. By and large, the more recent a technology is, the more of its costs aren't paid by the makers or the users of the technology but are pushed off onto someone else. This isn't accidental; as already noted, it's hardwired into the relationship between progress and market economics, and it bids fair to play a central role in the unraveling of the entire project of industrial civilization.

The same process of increasing externalities, though, has another face when seen from the point of view of the individual user of any given technology. When you externalize any cost of a technology, you become dependent on whoever or whatever picks up the cost you're not paying. What's more, you become dependent on the system that does the externalizing, and on whoever controls that system. Those dependencies aren't always obvious, but they impose costs of their own, some financial and some less tangible—and unlike the externalized costs, a great many of these secondary costs land directly on the user of the technology.

It's interesting, and may not be entirely accidental, that there's no commonly used term for the structure of externalities and dependencies that stand behind any technology. Such a term is necessary here, so we'll call the structure just named the technology's *externality system*. Given that turn of phrase, we can restate the point about progress made above: on the whole, the more recent a technology is, the larger the externality system on which it depends.

An example will be useful here, so let's compare the externality systems of a bicycle and an automobile. Like most other externality systems, these divide up into three categories: manufacture, maintenance, and use. Everything that goes into fabricating steel parts, for instance, all the way back to the iron ore in the mine, is an externality of manufacture; everything that goes into making lubricating oil, all the way back to drilling for the oil well, is an externality of maintenance; everything that goes into building roads suitable for bikes and cars is an externality of use.

The externality systems of bicycles and automobiles are both complex and include a great many things that aren't obvious at first glance. The crucial point, though, is that the car's externality system is far and away the more complex of the two. In fact, the bike's externality system is a subset of the car's, and this reflects the historical order in which the two technologies were developed. When the technologies that were needed for a bicycle's externality system came into use, bicycles appeared; when all the additional technologies needed for a car's externality system were added onto that foundation, cars followed. That sort of incremental addition of externality-generating technologies is by far the most common way that technology progresses.

We can thus restate the pattern just analyzed in a way that brings out some of its less visible and more troublesome aspects: by and large, each new generation of technology imposes more dependencies on its users than the generation it replaces. Again, a comparison between bicycles and automobiles will help make that clear. If you want to ride a bike, you've committed yourself to dependence on all the technical, economic, and social systems that go into manufacturing, maintaining, and using the bike; you can't own, maintain, and ride a bike without the steel mills that produce the frame, the chemical plants that produce the oil you squirt on the gears, the gravel pits that provide raw material for roads and bike paths, and so on.

On the other hand, you're not dependent on a galaxy of other systems that provide the externality system for your neighbor who drives. You don't depend on the immense network of pipelines, tanker trucks, and gas stations that provide him with fuel, or on the military and political arrangements that keep a disproportionate share of the world's petroleum flowing to North American consumers; you don't depend on the interstate highway system or the immense infrastructure that supports it; if you do the sensible thing and buy a bike that was made by a local craftsperson, your dependence on vast multinational corporations and all of their infrastructure, from sweatshop labor in Third World countries to financial shenanigans on Wall Street, is considerably smaller than that of your driving neighbor. Every dependency you have, your neighbor also has, but not vice versa.

The Downsides of Dependency

Whether or not these dependencies matter is a complex thing. Obviously there's a personal equation—some people like to be independent, while others are fine with being just one more cog in the megamachine—but there's also a historical factor to consider. In an age of economic expansion, the benefits of dependency very often outweigh the costs. In such periods, standards of living are rising, opportunities abound, and it's easy to offset the costs of any given dependency. In a stable economy, one that's neither growing nor contracting, the benefits and costs of any given dependency need to be weighed carefully on a case-by-case basis, as one dependency may be worth accepting while another costs more than it's worth. On the other hand, in an age of contraction and decline—or, shall we say, negative growth?—most dependencies are problematic, and some are lethal.

By many measures, the US economy has been contracting since before the bursting of the housing bubble in 2008; by some—in particular, the median and modal standards of living—it's been contracting since the 1970s, and the unmistakable hissing sound as air leaks out of the latest round of financial bubbles should be considered fair warning that another round of contraction is on its way.

With that in mind, it's time to talk about the downsides of dependency.

First of all, *dependency is expensive*. In the struggle for shares of a shrinking pie in a contracting economy, turning any available dependency into a cash cow is an obvious strategy, and one that's already very much in play. In a contracting economy, as everyone scrambles to hold onto as much as possible of the lifestyles of a more prosperous age, your profit is by definition someone else's loss, and dependency is just another weapon in the Hobbesian war of all against all.

Consider the conversion of freeways into toll roads, an increasingly popular strategy in large parts of the United States. Consider, for that matter, the soaring price of health care in the US, which hasn't been accompanied by any noticeable increase in quality of care or treatment outcomes. In the dog-eat-dog world of economic contraction, commuters and sick people are just two of many captive populations whose dependencies make them vulnerable to exploitation. As the spiral of decline continues, it's safe to assume that any dependency that can be exploited will be exploited, and the more dependencies you have, the more likely you are to be squeezed dry.

The same principle applies to power as well as money, and so *whoever owns the systems on which you depend, owns you*. In the United States, again, laws meant to protect employees from abusive behavior on the part of employers are increasingly ignored. As the number of the permanently unemployed keeps climbing year after year, employers know that those who still have jobs are desperate to keep them and will put up with almost anything in order to keep that paycheck coming in. The old adage about the inadvisability of trying to fight City Hall has its roots in this same phenomenon; no matter what rights you have on paper, you're not likely to get far with them when the other side can stop picking up your garbage and then fine you for creating a public nuisance, or engage in some other equally creative use of their official prerogatives. As decline accelerates, expect to see dependencies increasingly used as levers for exerting various kinds of economic, political, and social power at your expense.

Finally, and crucially, if you're dependent on a failing system, *when the system goes down, so do you.* That's not just an issue for the future. It's a huge, if still largely unmentioned, reality of life in today's United States and in most other corners of the industrial world as well. Most of today's permanently unemployed got that way because the job on which they depended for their livelihood got offshored or automated out of existence; much of the rising tide of poverty across the United States is a direct result of the collapse of political and social systems that once countered the free market's innate tendency to drive the gap between rich and poor to Dickensian extremes. For that matter, how many people who never learned how to read a road map are already finding themselves in random places far from help because something went wrong with their GPS units?

It's very popular among those who recognize the problem with being shackled to a collapsing system to insist that it's a problem for the future, not the present. They grant that dependency is going to be a losing bet someday, but everything's fine for now, so why not enjoy the latest technological gimmickry while it's here? Of course, that presupposes that you enjoy the latest technological gimmickry, which isn't necessarily a safe bet, and it also ignores the first two difficulties with dependency outlined above, which are very much present and accounted for right now. We'll let both those issues pass for the moment, though, because there's another factor that needs to be included in the calculation.

A practical example, again, will be useful here. In my experience, it takes about five years of hard work, study, and learning from your mistakes to become a competent vegetable gardener.

If you're transitioning from buying all your vegetables at the grocery store to growing them in your backyard, in other words, you need to start gardening about five years before your last trip to the grocery store. The skill and hard work that goes into growing vegetables is one of many things that most people in the world's industrial nations externalize, and those things don't just pop back to you when you leave the produce section of the store for the last time. There's a learning curve that has to be undergone.

Not that long ago, there used to be a subset of preppers who grasped the fact that stashing cartridges and canned wieners in a locked box at their deer camp cabin wasn't going to get them through the downfall of industrial civilization. Businesses targeting the prepper market thus used to sell them garden-in-a-box kits, which had seed packets for vegetables, a few tools, and a little manual on gardening. The one difficulty was that neither the preppers nor their suppliers had factored in the learning curve.

It's a good thing that Y2K, 2012, and all those other dates when instant doom was supposed to arrive turned out to be wrong, because I met a fair number of people who thought that having one of those kits would save them, even though they last grew a plant from seed in fourth grade. If the apocalypse had actually arrived, survivors a few years later would have gotten used to a landscape scattered with empty garden-in-a-box kits, overgrown garden patches, and the skeletal remains of preppers who starved to death because the learning curve lasted longer than they did.

The same principle applies to every other set of skills that has been externalized by people in today's industrial society. You can adopt those skills now, while you still have time to get through the learning curve when there's an industrial society around to make up for the mistakes and failures that are inseparable from learning; or you can try to adopt them later, when those same inevitable mistakes and failures will land you in a world of hurt. You can also adopt them now, when your dependencies haven't yet been used to empty your wallet and control your behavior; or you can try to adopt them later, when a much larger fraction of the resources and autonomy you might have used for the purpose will have been extracted from you by way of those same dependencies.

This applies with particular force to negative progress—that is, to the deliberate adoption of older, simpler, more independent technologies in place of the latest, dependency-laden offerings from the corporate machine. As decline becomes an inescapable fact of life in post-progress North America, decreasing dependence on externality systems is going to be an essential tactic.

Those who become early adopters of the retro future will have at least two, and potentially three, significant advantages. The first, as already noted, is that they'll be much further along the learning curve by the time rising costs, increasing instabilities, and cascading systems failures either put the complex technosystems out of reach or push the relationship between costs and benefits well over into losing-proposition territory. The second is that as more people catch onto the advantages of older, simpler, more sustainable technologies, surviving specimens of those technologies will become harder to find and more expensive to buy. In this case, as in many others, collapsing ahead of the rush is, among other things, the more affordable option.

The third advantage? Depending on exactly which old technologies you happen to adopt, and whether or not you have any talent for basement-workshop manufacture and the like, you may find yourself on the way to a viable new career, just when most other people will be losing their jobs—and their shirts. As the global economy comes unraveled and people in North America lose their current access to shoddy imports from Third World sweatshops, there will be a demand for a wide range of tools and simple technologies that still make sense in a deindustrializing world. Those who already know how to use such technologies will be prepared to teach others how to use them; those who know how to repair, recondition, or manufacture those technologies will have something to barter, or to sell for whatever form of currency happens to replace today's mostly hallucinatory forms of money.

That said, there's a barrier on the road to those productive outcomes. Like most of the barriers that have to be overcome in the process of facing the future, it has much more to do with ideas and assumptions than with hard physical realities. The barrier in question is one of the most rigidly held taboos in contemporary culture, and it has to be confronted in order to make sense of the project of deliberate technological regression.

4

THE HERESY OF TECHNOLOGICAL CHOICE

PORT TOWNSEND is a pleasant town in Washington State, perched on a bluff above the western shores of Puget Sound. Due to the vagaries of the regional economy, it mostly got bypassed by the twentieth century, and much of the housing stock dates from the Victorian era. It so happens that one couple who live there find Victorian technology, clothing, and personal habits more to their taste than the current fashions in these things, and they live, as thoroughly as they can, a Victorian lifestyle. The wife of the couple, Sarah Chrisman, wrote a book about her experiences[1] and got her canonical fifteen minutes of fame on the internet and the media as a result.

You might think, dear reader, that the people of Port Townsend would treat this as merely a harmless eccentricity, or even find it pleasantly amusing to have a couple in Victorian cycling clothes riding their penny-farthing bicycles on the city streets. To some extent, you'd be right, but it's the exceptions that I want to discuss here. Ever since they adopted their Victorian lifestyle, the Chrismans have been on the receiving end of constant harassment by people who find their presence in the community intolerable.[2] The shouted insults, the in-your-face confrontations, the death threats—they've seen it all. What's more, the appearance of Sarah Chrisman's book and various online articles related to it fielded, in response, an impressive flurry of spluttering online denunciations, which insisted among other things that the fact that she prefers to wear long skirts and corsets somehow makes her personally responsible for all the sins that have ever been imputed to the Victorian era.

Why? Why the fury, the brutality, and the frankly irrational denunciations directed at a couple whose lifestyle choices have got to count among the world's most harmless hobbies?

The reason's actually very simple. Sarah Chrisman and her husband have transgressed one of our culture's most rigidly enforced taboos. They've shown in the most irrefutable way, by personal example, that the technologies each of us use in our own lives are a matter of individual choice.

You're not supposed to say that in today's world. You're not even supposed to think it. You're allowed, at most, to talk nostalgically about how much more pleasant it must have been not to be constantly harassed and annoyed by the current round of officially prescribed technologies and to be squashed into the Procrustean bed of the narrow range of acceptable lifestyles that go with them. Even that's chancy in many circles these days and risks fielding a diatribe from somebody who just has to tell you, at great length and with obvious irritation, all about the horrible things you'd supposedly suffer if you didn't have the current round of officially prescribed technologies constantly harassing and annoying you.

The wistfulness in question doesn't have to be oriented toward the past. I long ago lost track of the number of people I've heard talk wistfully about what I've called the Ecotopian future, the default vision of a green tomorrow that fills most minds on the leftward end of the political and cultural spectrum. Unless you've been hiding under a rock for the past forty years, you already know every detail of the Ecotopian future: it's the place where wind turbines and solar panels power everything, everyone commutes by bicycle from their earth-sheltered suburban homes to their LEED–certified urban workplaces, everything is recycled, and social problems have all been solved because everybody, without exception, has come to embrace the ideas and attitudes currently found among upper-middle-class San Francisco liberals.

It's far from rare, at sustainability-oriented events, to hear well-to-do attendees waxing rhapsodically about how great life will be when the Ecotopian future arrives. If you encounter someone engaging in that sort of wistful exercise, and are minded to be cruel, ask the person who's doing it whether he (it's usually a man) bicycles to work, and if not, why not. Odds are you'll get to hear any number of frantic excuses to explain why the lifestyle that everyone's going to love in the Ecotopian future is one that he can't possibly embrace today.

If you want a good hard look behind the excuses and evasions, ask him how he got to the sustainability-oriented event you're attending. Odds are that he drove his SUV, in which there were no other passengers. If you press him about that, you can expect to see the dark heart of privilege and rage that underlies his enthusiastic praise of an imaginary lifestyle that he would never, not even for a moment, think of adopting himself.

I wish I were joking about the rage. It so happens that I don't have a car, a television, or a cell phone, and I have zero interest in ever having any of these things. My defection from the officially prescribed technologies and the lifestyles that go with them isn't as immediately obvious as Sarah Chrisman's, so I don't take as much day-to-day harassment as she does. Still, it happens from time to time that somebody wants to know if I've seen this or that television program, and in the conversations that unfold from such questions it sometimes comes out that I don't have a television at all.

In the Appalachians, where I now live, that revelation rarely gets a hostile response, and it's fairly common for someone else to say, "Good for you," or something like that. A lot of people here are very poor, and thus have a certain detachment from technologies and lifestyles that they know perfectly well they will never be able to afford. Back when I lived in prosperous Left Coast towns, on the other hand, mentioning that I didn't own a television routinely meant that I'd get to hear a long and patronizing disquisition about how I really ought to run out and buy a TV so I could watch this, that, or the other really wonderful program, without which my life must be intolerably barren and incomplete.

Any disagreement on my part, furthermore, reliably brought out a variety of furiously angry responses that seemed weirdly unconnected with the simple fact that I didn't happen to use the same technologies they did. Nor did it matter how often I explained that the reason I don't watch television is simply that I don't enjoy it. It's not the programming I find unenjoyable, by the way, but the technology itself; I get bored very quickly with watching little colored images jerking about on a glass screen, no matter what the images happen to be.

That's another taboo, by the way. It's acceptable to grumble about what's on television, but the technology itself is sacrosanct. You're not allowed to criticize it, much less to talk about the biases, agendas, and simple annoyances hardwired into television as a technology. If you try to bring any of that up, people will insist that you're criticizing the programming; if you correct them, they'll ignore the correction and keep on talking as though the programs on TV are the only thing under discussion.

I'd encourage those of my readers who aren't blinded by the terror of intellectual heresy to think, and think hard, about the taboo against technological choice—the insistence that you cannot, may not, and must not make your own choices when it comes to technology but must do as you're told and accept whatever technology the consumer society hands you, no matter how dysfunctional, harmful, or boring it turns out to be. That taboo is very deeply ingrained, far more potent than the handful of relatively weak taboos our society still applies to such things as sex and death, and most of the people you know obey it so unthinkingly that they never notice how it shapes their behavior. You may not notice how it shapes your own behavior, for that matter. The best way to see this for yourself is to pick a technology that annoys, harms, or bores you, but that you use anyway, and get rid of it.

Those who take that unthinkable step, and embrace the heresy of technological choice, are on the cutting edge of the future. In a world of declining resource availability, unraveling economic systems, and destabilizing environments, Sarah Chrisman and the many other people who make similar choices—there are quite a few of them these days, and more of them with each year that passes—are making a wise choice. By taking up technologies and lifeways from less extravagant eras, they're decreasing their environmental footprints and their vulnerability to faltering global technostructures, and they're also contributing to one of the crucial tasks of our age: the rediscovery of ways of being human that don't depend on levels of resource and energy consumption that can't be sustained for much longer.

Technobullying and Technoshaming

Yet the pressure toward technological conformity is a real force in today's society, and it can be measured in the collision between those who decide not to embrace the latest technologies and those who react to that decision. Today, a significant number of Americans, for one good reason or another, choose not to use cell phones, televisions, automobiles, microwave ovens, and an assortment of other currently fashionable technologies. On the other hand, an even larger number of Americans get really remarkably freaky about people who make such choices.

Some of the stories I've heard from readers of my blog are classic. There was the couple who don't enjoy television, and so don't own one, and had a relative ask them every single year, over and over again, if she could buy them a television for Christmas. They said no thank you every single year, and finally she went out and bought them a television anyway because she just couldn't stand the thought that they weren't watching one. There was the coworker who plopped a laptop playing some sitcom or other right down on the lap of one of my readers and demanded that the reader watch it, right then and there, so that they would have something to talk about. There was the person who, offended by another reader's lack of interest in television, finally shouted, "You must be living in a dream world!" Er, which of these people was spending four to six hours a day watching paid actors playing imaginary characters act out fictional events in contrived settings?

Televisions were far from the only focus of this sort of technobullying. Other readers reported getting similar reactions from other people because they didn't happen to have, and weren't interested in having, microwave ovens, smartphones, and so on down the list of currently fashionable technologies. The stories are really quite eye-opening, and not in a good way. Forget about all the popular cant that insists that you're free to make your own choices and have whatever lifestyle you want. According to a significant fraction of people in North America—and to judge from what my readers reported, that fraction isn't limited to any one class, income level, or region of the continent—the only freedom you're supposed to exercise, when it comes to technology, is that of choosing which brand label will be slapped on each item in the officially approved list of devices you're expected to own.

The prevalence of technobullying and technoshaming in today's society is a fascinating point, but there's another point, which is that there's a significant amount of pushback against it. By that I don't mean that there's some sort of groundswell of renunciation, leading people to walk away from technologies in the same spirit that led medieval ascetics to don hair shirts and flog themselves for the good of their souls. That's a common stereotype directed at those of us who aren't interested in the latest technologies, and it completely misses what's actually going on.

I'll use myself as an example here. As already noted, I don't own a television—I haven't owned one in my adult life—and as already noted, it's not because I have some moral or political objection to televisions or because I'm into self-denial or what have you. I don't own a television because I find watching television dull. To me, quite simply, the activity of watching little colored shapes jerk around on a screen is boring and irritating, not relaxing and enjoyable, no matter what the little colored shapes are supposed to be doing.

Yes, I grew up with a television in the house. I experienced plenty of it back in the day, and I have no interest in experiencing any more, because I don't like it. It really is that simple. It's just as simple for others as well: they don't find this or that technology enjoyable, useful, or relevant to their lifestyles, and so they've chosen to do something else with their money and time. Shouldn't so simple and personal a choice be their own business and nobody else's? To judge by the reactions commonly fielded by those who make such choices, apparently not.

The arguments that generally get used by technobullies and technoshamers, in their attempts to argue against this expression of personal freedom, involve a decidedly odd use of logic. One of my favorites, which I've encountered repeatedly, is the insistence that any lack of enthusiasm for a fashionable technology obviously means that the heretic is conspiring to deprive everyone else of that technology. It's hard to see how that makes any kind of sense, but there it is; when it comes to television, in particular, you can count on being told that if you don't have one, obviously you must be trying to deny it to everyone.

Another common bit of rhetoric is the insistence that people who don't have televisions, microwaves, cell phones, or what have you, are hypocrites if they have internet connections. I suppose a case could be made that if a lack of interest in having a television, a microwave, or a cell phone was motivated by a passion for hair-shirt asceticism, having an internet connection could be interpreted as the kind of slacking that used to get you thrown out of the really top-notch hermitages. From any other perspective, it's a triumph of absurdity. If people are in fact allowed to choose, from among the available technologies, those that make them happiest—as the cheerleaders of the consumer economy delight to insist—what could be wrong with choosing some old technologies and some newer ones, if that's the mix you prefer?

Then there's the claim that if you return to an older technology, you have to take the social practices and cultural mores of its heyday as well. Bizarre as this claim is, it's astonishingly common; it's played a very large role in the venom flung at the Chrismans, among many others. Suggest that it makes sense to return to the technologies of the 1950s, say, and I can promise you that critics will insist that this is tantamount to approving of or trying to bringing back the social customs of the 1950s.

What makes this paralogic fascinating is that people in 1950, say, didn't have a single set of social customs. Even in the United States, a drive from Greenwich Village to rural Pennsylvania in that year would have met with remarkable cultural diversity among people using the same technology. The point could be made even more strongly by noting that roughly the same technological suite was in use that year in Paris, Djakarta, Buenos Aires, Tokyo, Tangiers, Novosibirsk, Guadalajara, and Lagos, and the social customs of 1950s middle America didn't follow the technology around to these distinctly diverse settings. Pointing that out, though, has no result on the believers.

Along the same lines is the rhetoric that consists of endlessly rehashing whatever bad things happened in the past, even when the bad things in question had nothing to do with the technology and vice versa. People like the Chrismans, who prefer Victorian furnishings and clothing to their modern equivalents, field this bizarre non sequitur all the time. Here again, there's some heavy-duty illogic involved. If a technology that was invented and used in the 1850s, say, is permanently tarred with the various social evils of that era, and ought to be rejected because those evils happened, wouldn't that also mean that the internet is just as indelibly tarred with the social evils of the modern era, and ought to be discarded because bad things are happening in the world today? What's sauce for the goose, after all, is sauce for the gander.

Finally, there's the capstone of the whole edifice of unreason, the insistence that anybody who doesn't use the latest, hottest technology wants to go "back to the caves," or to even take all of humanity to that much-denounced destination. "The caves" have a bizarre gravitational effect on the imagination of a certain class of modern thinkers. Everything that's not sufficiently modern and up to date, in their minds, somehow morphs into the bearskin kilts and wooden clubs that so many of us still, despite well over a century of detailed archeological evidence, insist on pushing onto our prehistoric ancestors.[3]

When people of this kind archly dismiss people like the Chrismans as going "back to the caves," they're engaged in a very interesting kind of absurdity. Do cavemen and Victorians belong on the same level? Sure, cavemen had flush toilets and central heating, daily newspapers and public libraries, not to mention factories, railways, global maritime trade, a telegraph network covering much of the planet's land surface, and a great deal more of the same kind! That's absurd, of course. It's even more absurd to insist that people who simply don't enjoy using this or that technology, and so don't use it, are going back to "the caves"—but I can promise you, dear reader, from my own personal experience, that if you show a lack of interest in any piece of fashionable technology, you'll have this phrase thrown at you.

That happens because "the caves" aren't real. They aren't, for example, the actual cave-shrines of the Magdalenian people who lived in Western Europe fifteen thousand years ago, whose lifestyles were quite similar to those of Native Americans before Columbus, and who used to go deep into the caves of Europe to paint sacred images that still stun the viewer today by their beauty and artistry. "The caves" of contemporary rhetoric, rather, are thoughtstopping abstractions, bits of verbal noise that people have been taught to use so they don't ask inconvenient questions about where this thing called "progress" is taking us and whether any sane person would actually want to go there. Flattening out the entire complex richness of the human past into a single cardboard bogeyman labeled "the caves" is one way to do that. So is papering over the distinctly ugly future we're making for ourselves with a screen shot or two from a Jetsons cartoon and a gaudy banner saying "We're headed for the stars!"

The Shadows in the Cave

It's really rather fascinating, all things considered, that the image of the cave should have been picked up for that dubious purpose. Not that long ago, most literate people in the Western world tended to have a very different image come to mind when someone mentioned caves. That was because of a man named Aristocles of Athens, who lived a little more than 2,300 years ago and whose very broad shoulders got him the nickname Plato. In the longest, most influential, and most problematic of his works, usually called *The Republic*—a bad choice, as this word nowadays has connotations of rights under law that the Greek title *Politeia* lacks—he framed his discussion of the gap between perception and reality with an arresting image.

Imagine, Plato says, that we are all shackled in a cave, unable to turn our heads to either side. All we can see are dark shapes that move this way and that on the flat wall of the cave in front of us. Those dark shapes are all we know. They are our reality.

Now imagine that one of these prisoners manages to get loose from his shackles, and turns away from the cave wall and the dark shapes on it. He's in for a shock, because what he sees when he turns around is a bonfire, and people moving objects in front of the flames so that the objects cast shadows on the cave wall.

Everything he thought was reality is simply a shadow cast by these moving objects.

If the prisoner who's gotten loose pays attention, furthermore, he might just notice that the contents of the cave aren't limited to the bonfire, the prisoners, the objects casting the shadows and the people who manipulate those objects. Off past the bonfire, on one side of the cave, the floor slopes upwards, and in the distance is a faint light that doesn't seem to come from the fire at all. If our escaped prisoner is brave enough, he might go investigate that light. As he does so, the bonfire and the shadows slip into the darkness behind him, and the light ahead grows brighter and clearer.

Then, if he's brave enough and keeps going, he steps out of the cave and into the sunlight. That's not an easy thing, either, because the light is so much more intense than the dim red glow in the cave that, for a while, he can't see a thing. He stumbles, rubs his eyes, tries to find his bearings, and discovers that the detailed knowledge he had of the way shadows moved on the cave wall won't help him at all in this new, blazingly bright realm. He has to discard everything he thinks he knows and learn the rules of an unfamiliar world.

Bit by bit, though, he accomplishes this. His eyes adapt to the sunlight; he learns to recognize objects and to sense things—color, for example, and depth—which didn't exist in the shadow-world he thought he inhabited when he was still a prisoner in the cave. Eventually he can even see the Sun and know where the light that illumines the real world actually comes from.

Now, Plato says, imagine that he decides to go back into the cave to tell the remaining prisoners what he's seen. To begin with, it's going to be rough going, because his eyes have adapted to the brilliant daylight and so he's going to trip and stumble on the way down. Once he gets there, anything he says to the prisoners is going to be dismissed as the most consummate rubbish: what is this nonsense about color and depth, and a big bright glowing thing that crosses something called the sky? What's more, the people to whom he's addressing his words are going to misunderstand them, thinking that they're about the shadow-world in front of their eyes—after all, that's the only reality they know—and they're going to decide that he must be an idiot because nothing he says has anything to do with the shadow-world.

Plato didn't mention that the prisoners might respond by trying to drag the escapee back into line with them and bully him into putting his shackles back on, though that's generally the way such things work out in practice. Plato also never saw a television, which is unfortunate in a way—if he had, he could have skipped the complicated setup with the bonfire and the people waving around objects that cast shadows, and simply said, "Imagine that we're all watching television in a dark room."

Now, of course, Plato had his own reasons for using the cave metaphor, and he developed it in directions that aren't relevant to the present discussion. The point I want to make here is that every technology is a filter that shapes the way we experience and interact with the world. In some cases, such as television, the filtering effect is so drastic it's hard not to see—unless, that is, you don't want to see it. In other cases, it's subtle. There are valid reasons people might want to use one filter rather than another, or to set aside an assortment of filters in order to get a clearer view of some part of the world or their lives.

There are also, as already noted, matters of personal choice. Some of us prefer sun and wind and depth and color to the play of shadows on the walls of the cave. That doesn't mean that we're going to drag those who don't share that preference out into the blinding light, or that we're going to turn ourselves into the Throg the Cave Man shadow that's being waved around so enthusiastically on one corner of the cave wall. It does mean that a significant number of people are losing interest in the shadow-play and are clambering up the awkward but rewarding journey into the sunlight and the clean cool air, and it may just mean as well that those who try to bully them into staying put and staring at shadows may have less success than they expect.

That said, the pressure to conform is real, and behind it lies a clear but unstated sense of the nature of our current predicament. Quite a few of the odder features of contemporary American culture, in fact, make perfect sense if you assume that everybody knows exactly what's wrong and what's coming as our society rushes, pedal to the metal, toward its face-first collision with the brick wall of the future. I've come to think, in fact, that our apparent blindness to the future is largely an illusion, and a deliberate one. It's not that people don't get it. They get it all too clearly, and they just wish that those of us on the fringes would quit reminding them of the imminent impact so they can spend whatever time they've got left in as close to a state of blissful indifference as they can possibly manage.

That realization first came to me while I was sitting in a restaurant, which I visited, not by my own choice, while carpooling down to Baltimore one summer day. I'll spare you the name of the place; it was one of those fashionable beer-and-burger joints where the waitresses have all been outfitted with skirts almost long enough to cover their underwear, bare midriffs, and the sort of push-up bras that made them look uncomfortably like inflatable dolls—an impression that their too obviously scripted jiggle-and-smile routines did nothing to dispel.

Still, that wasn't the thing that made the restaurant memorable. It was the fact that every wall in the place had television screens on it. By this I don't mean that there was one screen per wall; I mean that they were lined up side by side right next to each other, covering the upper part of every single wall in the place, so that you couldn't raise your eyes above head level without looking at one. They were all over the interior partitions of the place, too. There must have been forty of them in one not too large restaurant, each one blaring something different into the thick air while loud syrupy music spattered down on us from speakers on the ceiling and the waitresses smiled mirthlessly as they went through their routines. My burger and fries were tolerably good, and two tall glasses of Guinness will do much to ameliorate even so charmless a situation; still, I was glad to get back on the road.

It's worth pointing out that all this is quite recent. Not that many years ago, it was tolerably rare to see a TV screen in an American restaurant, and even those bars that had a television on the premises for the sake of football season generally had the grace to leave the thing off the rest of the time. Within the past decade, I've watched televisions sprout in restaurants and pubs I used to enjoy, for all the world like buboes on the body of a plague victim: first one screen, then several, then one on each wall, then metastasizing across the remaining space. Meanwhile, people who used to go to coffee shops and the like to read the papers, talk with other patrons, or do anything else you care to name are now sitting in the same coffee shops in total silence, hunched over their allegedly smart phones like so many scowling gargoyles on the walls of a medieval cathedral.

Yes, there were people in the restaurant crouched in the classic gargoyle pose over their allegedly smart phones, too, and that probably also had something to do with my realization that evening. It so happens that the evening before my Baltimore trip I'd recorded a podcast interview with Chris Martenson on his Peak Prosperity show,[4] and he'd described to me a curious response he'd been fielding from people who attended his talks on the end of the industrial age. He called it "the iPhone moment"—the point at which any number of people in the audience pulled that particular technological toy out of their jacket pockets and waved it at him, insisting that its mere existence somehow disproved everything he was saying.

As modern superstitions go, this one is pretty spectacular. Let's take a moment to look at it rationally. Do iPhones produce energy? Nope. Will they refill our rapidly depleting oil and gas wells, restock the ravaged oceans with fish, or restore the vanishing topsoil from the world's fields? Of course not. Will they suck carbon dioxide from the sky, get rid of the vast mats of floating plastic that clog the seas, or do something about the steadily increasing stockpiles of nuclear waste that are going to sicken and kill people for the next quarter of a million years unless the waste gets put someplace safe—if there is anywhere safe to put it at all? Not a chance. As a response to any of the predicaments that are driving the crisis of our age, iPhones are at best irrelevant. Since they consume energy and resources, and the sprawling technosystems that make them function consume energy and resources at a rate orders of magnitude greater, they're part of the problem, not any sort of a solution.

Of course, the people waving their iPhones at Chris Martenson aren't thinking about any of these things. A good case could be made that they're not actually thinking at all. Their reasoning, if it can be called that, is that the existence of iPhones proves that progress is still happening, and this somehow proves that progress will inevitably bail us out from the impacts of every one of the predicaments we face.

To call this magical thinking is an insult to honest sorcerers. It's another example of the arbitrary linkage of verbal noises to emotional reactions that all too often passes for thinking in today's America. Readers of classic science fiction may find all this weirdly reminiscent of a scene from some edgily updated version of H.G. Wells's *The Island of Doctor Moreau*: "Not to doubt Progress: that is the Law. Are we not Men?"

Seen from a different perspective, though, there's a definite if unmentionable logic to "the iPhone moment," and it has much in common with the metastatic spread of television screens across pubs and restaurants in recent years. These allegedly smart phones don't do anything to fix the rising spiral of problems besetting industrial civilization, but they make it easier for people to distract themselves from those problems for a little while longer.

That, I'd like to suggest, is also what's driving the metastasis of television screens in the places that people used to go to enjoy a meal, a beer, or a cup of coffee and each other's company. These days, that latter's too risky. Somebody might mention a friend who lost his job and can't get another one, a spouse who gets sicker with each overpriced and ineffective prescription drug the medical industry pushes on her, a kid who didn't come back from Afghanistan, or the like, and then it's right back to the reality that everyone's trying to avoid. It's much easier to sit there in silence staring at little colored pictures on a glass screen, from which all such troubles have been excluded.

The Burden of Denial

Of course, that habit has its own downsides. To begin with, those who are busy staring at the screens have to know, on some level, that sooner or later it's going to be their turn to lose their jobs, or have their health permanently wrecked by the side effects their doctors didn't get around to mentioning, or have their kids fail to come back from the United States' latest war of choice, or the like. That's why so many people these days put so much effort into insisting that the poor and vulnerable are to blame for their plight. The people who say this know perfectly well that it's not true, but repeating such claims over and over again is the only defense they've got against the bitter awareness that their jobs, their health, and their lives or those of the people they care about could all too easily be next on the chopping block.

What makes this all the more difficult for most Americans to face is that none of these events are happening in a vacuum. They're part of a broader process, the decline of modern industrial society in general and the United States of America in particular. Outside the narrowing circles of the well-to-do, standards of living for most Americans have been declining since the 1970s, along with standards of education, public health, and most of the other things that make for a prosperous and stable society. Today, a nation that once put human bootprints on the Moon can't afford to maintain its roads and bridges or keep its cities from falling into ruin. Hiding from that reality in an imaginary world projected onto glass screens may be comforting in the short term; the mere fact that realities don't go away just because they're ignored does nothing to make this choice any less tempting.

What's more, the world into which that broader process of decline is bringing us is not one in which staring at little colored pictures on a glass screen will count for much. Quite the contrary, it promises to be a world in which raw survival, among other things, will depend on having achieved at least a basic mastery of one or more of a very different range of skills. There's no particular mystery about those latter skills; they were, in point of fact, the standard set of basic human survival skills for thousands of years before those glass screens were invented, and they'll still be in common use when the last of the glass screens has weathered away into sand; but they have to be learned and practiced before they're needed, and there may not be all that much time left to learn and practice them before hard necessity comes knocking at the door.

I think a great many people who claim that everything's fine are perfectly aware of all this. They know what the score is; it's doing something about it that's the difficulty, because taking meaningful action at this very late stage of the game runs headlong into at least two massive obstacles. One of them is practical in nature, the other psychological, and human nature being what it is, the psychological dimension is far and away the most difficult of the two.

Let's deal with the practicalities first. We are all going to have to make do in the future with less energy and resources, and less of the products and services produced by the consumption of energy and resources. Using less of them now frees up time, money, and other resources that can be used to get ready for the inevitable transformations. It also makes for decreased dependence on systems and resources that in many cases are already beginning to fail, and in any case will not be there indefinitely in a future of hard limits and inevitable scarcities.

On the other hand, using less of anything flies in the face of two powerful forces in contemporary culture. The first is the ongoing barrage of advertising meant to convince people that they can't possibly be happy without the latest time-, energy-, and resource-wasting trinket that corporate interests want to push on them. The second is the stark shivering terror that seizes most Americans at the thought that anybody might think that they're poorer than they actually are. Americans like to think of themselves as proud individualists, but like so many elements of the American self-image, that's an absurd fiction. These days, as a rule, Americans are meek conformists who tremble with horror at the thought that they might be caught straying in the least particular from whatever other people expect of them.

That's part of what lies behind the horrified response that comes up the moment someone suggests that using less might be a meaningful part of our response to the crises of our age. When people go around insisting that not buying into the latest overhyped and overpriced technology is tantamount to going back to the caves, you can tell that what's going on in their minds has nothing to do with the realities of the situation and everything to do with stark unreasoning fear. Point out that a mere thirty years ago, people got along just fine without email and the internet, and you're likely to get an even more frantic and abusive reaction, precisely because your listener knows you're right and can't deal with the implications.

This is where we get into the psychological dimension. What James Howard Kunstler has usefully termed the psychology of previous investment is a massive cultural force in today's industrial societies.[5] The predicaments we face today are in very large part the product of a long series of bad decisions that were made over the past four decades or so. Most people in the United States, even those who had little to do with making those decisions, enthusiastically applauded them and treated those who didn't applaud with no small amount of abuse and contempt. Admitting just how misguided those decisions turned out to be thus requires a willingness to eat crow that isn't common these days. Thus there's a strong temptation to double down on the bad decisions, wave those iPhones in the air, and put a few more television screens on the walls to keep the cognitive dissonance at bay for a little while longer.

That temptation isn't an abstract thing. It rises out of the raw emotional anguish woven throughout our collective attempt to avoid looking at the future we've made for ourselves. The intensity of that anguish can be measured most precisely in one small but telling point: the number of people whose final response to the lengthening shadow of the future is, "I hope I'll be dead before it happens."

Think about those words for a moment. It used to be absolutely standard, and not only in North America, for people of every social class below the very rich to work hard, save money, and do without so that their children could have a better life than they had. That parents could say to their own children, "I got mine, Jack; too bad your lives are going to suck," belonged in the pages of lurid dime novels, not in everyday life.

Yet that's exactly what the words "I hope I'll be dead before it happens" imply. The destiny that's overtaking the industrial world isn't something imposed from outside; it's not an act of God or nature or callous fate; rather, it's unfolding with mathematical exactness from the behavior of those who benefit from the existing order of things. It could be ameliorated significantly if those same people were to let go of the absurd extravagance that passes for a normal life in the modern industrial world these days. It's just that the act of letting go involves an emotional price that few people are willing to pay.

Thus I don't think that anyone says "I hope I'll be dead before it happens" lightly. I don't think the people who are consigning their own children and grandchildren to a ghastly future, and placing their last scrap of hope on the prospect that they themselves won't live to see that future arrive, are making that choice out of heartlessness or malice. The frantic concentration on glass screens, the bizarre attempts to banish unwelcome realities by waving iPhones in their faces, and the other weird behavior patterns that surround modern industrial society's nonresponse to its impending future are signs of the enormous strain that so many Americans these days are under as they try to keep pretending—in the teeth of the facts—that nothing is wrong.

Denying a reality that's staring you in the face is an immensely stressful process, and the stress gets worse as the number of things that have to be excluded from awareness mounts up. These days, that list is getting increasingly long. Look away from the pictures on the glass screens, and the United States is visibly a nation in rapid decline: its cities collapsing, its infrastructure succumbing to decades of malign neglect, its politics mired in corruption and permanent gridlock, its society frayed to breaking, and the natural systems that support its existence passing one tipping point after another and lurching through chaotic transitions. It's one useful measure of our predicament that Oklahoma has passed California as the most seismically active state in the Union, as countless gallons of fracking fluid pumped into deep disposal wells remind us that nothing ever really "goes away."[6] It's no wonder that so many shrill voices these days are insisting that nothing is wrong, or that it's all the fault of some scapegoat or other, or that Jesus or the Space Brothers or somebody will bail us out any day now, or that we're all going to be wiped out shortly by some colorful Hollywood cataclysm that, please note, is never our fault.

There is, of course, another option. Quite a few people have broken themselves out of the trap, or were popped out of it willy-nilly by some moment of experience just that little bit too forceful to yield to the exclusionary pressure. Some of them have talked to me about how the initial burst of terror—no, no, you can't say that, you can't *think* that!—gave way to an immense feeling of release and freedom, as the burden of keeping up the pretense dropped away and left them able to face the world in front of them at last.

I suspect, for what it's worth, that a great many more people are going to be passing through that transformative experience in the years immediately ahead. A majority? Almost certainly not. To judge by historical precedents, the worse things get, the more effort will go into the pretense that nothing is wrong at all, and the majority will cling like grim death to that pretense until it drags them under. That said, a large minority might make a different choice: to let go of the burden of denial soon enough to matter, to let themselves plunge through those moments of terror and freedom, and to haul themselves up, shaken but alive, onto the unfamiliar shores of the future.

5

SUSTAINABLE TECHNOLOGIES

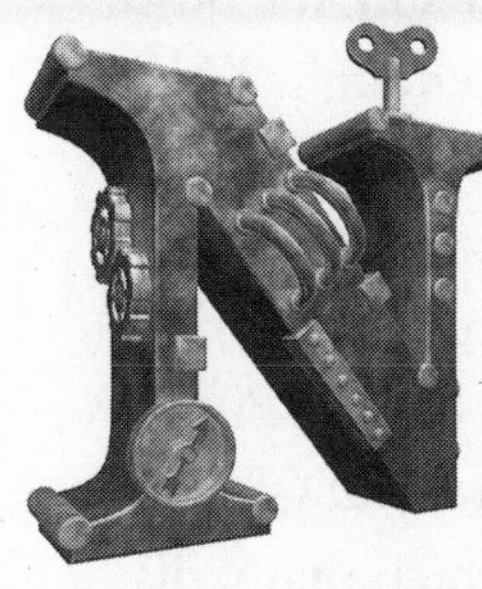

NOT LONG AGO a friend and I were talking about faith in progress, the ersatz religion that claims that all human history follows an ever-ascending arc from the caves to the stars. In the course of the discussion, my friend noted how disappointed he'd been with a book about the future that backed away from tomorrow's challenges into the shelter of a comforting thoughtstopper: "Technology will always be with us."

It's a very common utterance. Let's take a moment, though, to think about what, if anything, it actually means. Taken in the most literal sense, it's true but trivial. Toolmaking is one of our species' core evolutionary strategies, and so it's a safe bet that human beings will have some variety of technology or other as long as our species survives. That requirement could just as easily be satisfied, though, by a flint hand ax as by a laptop computer—and a flint hand ax is presumably not what people who use that particular thoughtstopper have in mind.

Perhaps we might rephrase the credo, then, as "Modern technology will always be with us." That's also true in a trivial sense, and false in another, slightly less trivial sense. In the first sense, every generation has its own modern technology; the latest up-to-date flint hand axes were, if you'll pardon the pun, cutting-edge technology in the time of the Neanderthals. In the second sense, much of every generation's modern technology goes away promptly with that generation; whichever way the future goes, much of what counts as modern technology today will soon be no more modern and cutting edge than eight-track tape players or Victorian magic-lantern projectors. That's as true if we get a future of continued progress as it is if we get a future of regression and decline.

Perhaps our author means something like "Technological complexity never decreases." This is less trivial but it's quite simply false, as historical parallels show clearly enough. Much of the technology of the Roman era, from wheel-thrown pottery to central heating, was lost in most of the Western Empire and had to be brought in from elsewhere centuries later.[1] In the dark ages that followed the fall of Mycenaean Greece, even so simple a trick as the art of writing was lost, while the history of Chinese technology before the modern era is a cycle in which many discoveries made during the heyday of each great dynasty were lost in the dark age that followed its decline and fall, and had to be rediscovered when stability and prosperity returned. For people living in each of these troubled times, technological complexity did in fact decrease.

For that matter, who is the "us" that technology is always supposed to be with? Many people on Earth right now have no access to the technologies that middle-class Americans take for granted. For all the good that modern technology does them, today's rural subsistence farmers, laborers in sweatshop factories, and the like might as well be living in some earlier era. I suspect our author is not thinking about such people, though, and the credo thus might be phrased as "Some technology at least as complex as what middle-class people in the industrial world have now, providing the same services they have come to expect, will always be available to people of that same class."

Depending on how you define social classes, that's either true but trivial—if "being middle class" equals "having access to the technology today's middle classes have," no middle-class people will ever be deprived of such a technology because, by definition, there will be no middle-class people once the technology stops being available—or nontrivial but clearly false—plenty of people who think of themselves as middle-class Americans right now are losing access to a great deal of technology as economic contraction deprives them of their jobs and incomes and launches them on new careers of downward mobility and radical impoverishment.

Well before the analysis got this far, of course, anyone who's likely to mutter the credo "Technology will always be with us" will have jumped up and yelled, "Oh, for heaven's sake, you know perfectly well what I mean when I use that word! You know, *technology!*"—or words to that effect. Now of course, I do know exactly what the word means in that context: it's a vague abstraction with no real conceptual meaning at all but an ample supply of raw emotional force. Like other thoughtstoppers of the same kind, it serves as a verbal bludgeon to prevent people from talking or even thinking about the brittle, fractious, ambivalent realities that shape our lives these days. Still, let's go a little further with the process of analysis, because it leads somewhere that's far from trivial.

Keep asking a believer in the credo we're discussing the sort of annoying questions I've suggested above, and if the conversation lasts long enough, sooner or later you're likely to get a redefinition that goes something like this: "The coming of the industrial revolution was a major watershed in human history, and no future society of any importance will ever again be deprived of the possibilities opened up by that revolution." Whether or not that turns out to be true is a question nobody today can answer, but it's a claim worth considering, because history shows that enduring shifts of this kind do happen from time to time.

The agricultural revolution of c. 9000 BCE and the urban revolution of c. 3500 BCE were both decisive changes in human history. Even though there were plenty of nonagricultural societies after the first, and plenty of non-urban societies after the second, the possibilities opened up by each revolution were always options thereafter, when and where ecological and social circumstances permitted. Some 5,500 years passed between the agricultural revolution and the urban revolution, and since it's been right around 5,500 years since the urban revolution began, a case could probably be made that we were due for another. Still, let's take a closer look at the putative third revolution. What exactly *was* the industrial revolution? What changed, and what future awaits those changes?

The Four Industrial Revolutions

That's a far more subtle question than it might seem at first glance, because the cascade of changes that fit under the very broad label "the industrial revolution" weren't all of a piece. I'd like to suggest, in fact, that there was not one industrial revolution but four of them. Lewis Mumford's important 1934 study *Technics and Civilization* identified three of those revolutions, though the labels he used for them—the eotechnic, paleotechnic, and neotechnic phases—shoved them into a linear scheme of progress that distorts many of their key features. Instead, I propose to borrow the same habit people use when they talk about the Copernican and Darwinian revolutions, and name the revolutions after individuals who played crucial roles in making them happen.

First of all, then—corresponding to Mumford's eotechnic phase—is the Baconian revolution, which got underway around 1600. It takes its name from Francis Bacon, who was the first significant European thinker to propose that what he called natural philosophy and we call science ought to be reoriented away from the abstract contemplation of the cosmos and toward making practical improvements in the technologies of the time.

Such improvements were already underway, carried out by a new class of "mechanicks," who had begun to learn by experience that building a faster ship, a sturdier plow, a better spinning wheel, or the like could be a quick route to prosperity, and encouraged by governments eager to leverage new inventions into the more valued coinage of national wealth and military victory.

The Baconian revolution, like those that followed it, brought with it a specific suite of technologies. Square-rigged wooden ships capable of long deepwater voyages revolutionized international trade and naval warfare; canals and canal boats had a similar impact on domestic transport systems. New information and communication media—newspapers, magazines, and public libraries—were crucial elements of the Baconian technological suite, which also encompassed major improvements in agriculture and in metal and glass manufacture and significant developments in the use of wind and water power as well as the first factories using division of labor to allow mass production.

The second revolution—corresponding to Mumford's paleotechnic phase—was the Wattean revolution, which got started around 1780. This takes its name, of course, from James Watt, whose redesign of the steam engine turned it from a convenience for the mining industry to the throbbing heart of a wholly new technological regime, replacing renewable energy sources with concentrated fossil fuel energy and putting that latter to work in every economically viable setting. The steamship was the new vehicle of international trade, the railroad the corresponding domestic transport system; electricity came in with steam, and so did the telegraph, the major new communications technology of the era, while mass production of steel via the Bessemer process had a massive impact straight across every dimension of economics and technology.

The third revolution—corresponding to Mumford's neotechnic phase—was the Ottonian revolution, which took off around 1890. I've named this revolution after Nikolaus Otto, who invented the four-cycle internal combustion engine in 1876 and kickstarted the process that turned petroleum from a convenient source of lamp fuel to the resource that brought the industrial age to its zenith. In the Ottonian era, international trade shifted to diesel-powered ships, supplemented later on by air travel; the domestic transport system was the automobile; the rise of vacuum-tube electronics made radio (including television, which is simply an application of radio technology) the major new communications technology; and the industrial use of organic chemistry, turning petroleum and other fossil fuels into feedstocks for plastics, gave the Ottonian era its most distinctive materials.

The fourth revolution, which hadn't yet begun when Mumford wrote his book, was the Fermian revolution, which can be dated quite precisely to 1942 and is named after Enrico Fermi, the designer and builder of the first successful nuclear reactor. The keynote of the Fermian era was the application of subatomic physics, not only in nuclear power but also in solid-state electronic devices such as the transistor and the photovoltaic cell. In the middle years of the twentieth century, a great many people took it for granted that the Fermian revolution would follow the same trajectory as its Wattean and Ottonian predecessors: nuclear fission power would replace diesel power in freighters; electricity would elbow aside gasoline as the power source for domestic transport; and nucleonics would become as important as electronics as a core element in new technologies yet unimagined.

Unfortunately for those expectations, nuclear fission power turned out to be a technical triumph but an economic flop. Claims that fission plants would make electricity too cheap to meter ran face first into the hard fact that no nation anywhere has been able to support a nuclear power industry without huge and ongoing government subsidies; and nuclear-powered ships were tried out for commercial uses, found hopelessly uneconomical, and thereafter relegated to naval vessels, which didn't have to turn a profit and so could afford to ignore the sky-high construction and operating costs. Nucleonics turned out to have certain applications but nothing like as many or as lucrative as the giddy forecasts of 1950 suggested. Solid-state electronics, on the other hand, turned out to be economically viable, at least in a world with ample fossil fuel supplies, and made the computer and the era's distinctive communications medium, the internet, economically viable propositions.

The Wattean, Ottonian, and Fermian revolutions thus had a core theme in common. Each of them relied on a previously untapped energy resource—coal, petroleum, and uranium, respectively—and set out to build a suite of technologies to exploit that resource and the forms of energy it made available. The scientific and engineering know-how that was required to manage each power source then became the key toolkit for the technological suite that unfolded from it; from the coal furnace, the Bessemer process for making steel was a logical extension, just as the knowledge of hydrocarbon chemistry needed for petroleum refining became the basis for plastics and the chemical industry, and the same revolution in physics that made nuclear fission reactors possible also launched solid-state electronics; it's not often remembered, for example, that Albert Einstein got his Nobel Prize for understanding the process that makes photovoltaic cells work, not for the theory of relativity.

The core technologies of the Wattean, Ottonian, and Fermian eras thus all depended on access to large amounts of specific nonrenewable resources. Fermian technology, for example, demands fissible material for its reactors and rare earth elements for its electronics, among many other things. Ottonian technology demands petroleum, natural gas, and an assortment of metal ores. Wattean technology demands coal and iron ore. It's sometimes possible to substitute one set of materials for another—say, to process coal into liquid fuel—but there's always a major economic cost involved, even if there's an ample and inexpensive supply of the other resource that isn't needed for some other purpose.

In today's world, by contrast, the resources needed for all three technological suites are being used at breakneck rates, and thus are either already facing depletion or will do so in the near future. When coal has already been mined so heavily that sulfurous, low-energy brown coal—the kind that miners in the nineteenth century used to discard as waste—has become the standard fuel for coal-fired power plants, for example, it's a bit late to talk about a coal-to-liquids program to replace any serious fraction of the world's petroleum consumption: the attempt to do so would send coal prices soaring to economy-wrecking heights. Richard Heinberg has discussed at length in his useful book *Peak Everything*, for that matter, the fact that a great deal of the coal still remaining in the ground will take more energy to extract than it will produce when burned, making it an energy sink rather than an energy source.

Thus we can expect very large elements of Wattean, Ottonian, and Fermian technologies to stop being economically viable in the years ahead, as depletion drives up resource costs and the knock-on effects of the resulting economic contraction force down demand. That doesn't mean that every aspect of those technological suites will go away, to be sure. It's not at all unusual, in the wake of a fallen civilization, to find "orphan technologies" that once functioned as parts of a coherent technological suite still doing their jobs long after the rest of the suite has fallen out of use. Just as Roman aqueducts kept bringing water to cities in the post-Roman dark ages whose inhabitants had neither the resources nor the knowledge to build anything of the kind, it's quite likely that (say) hydroelectric facilities in certain locations will stay in use for centuries to come, powering whatever electrical equipment can be maintained or built from local resources, even if the people who tend the dams and use the electricity have long since lost the capacity to build turbines, generators, or dams at all.

The Baconian Legacy

Yet there's another issue involved, because the first of the four industrial revolutions I've discussed—the Baconian revolution—*was not dependent on nonrenewable resources*. The suite of technologies that unfolded from Francis Bacon's original project used the same energy sources that everyone in the world's urban-agricultural societies had been using for more than three thousand years: human and animal muscle, wind, water, and heat from burning biomass. Unlike the revolutions that followed it, to put the same issue in a different but equally relevant way, the Baconian revolution worked within the limits of the energy budget the Earth receives each year from the Sun, instead of drawing down stored sunlight from the Earth's store of fossil carbon or stored starlight from its much more limited store of fissible isotopes. Baconian technologies simply used that annual solar budget in a more systematic way than previous societies managed—by directing the intellectual skills of the natural philosophers of the day toward practical ends.

Because of their dependence on nonrenewable resources, the three later revolutions were guaranteed all along to be transitory phases. The Baconian revolution need not be, and I think that there's a noticeable chance that it will not be. By that I mean, to begin with, that the core intellectual leap that made the Baconian revolution possible—the scientific method—is sufficiently widespread at this point that, with a little help, it may well get through the decline and fall of our civilization. If it does so, it will likely become part of the standard toolkit of future civilizations in much the same way that classical logic survived the wreck of Rome to be taken up by successor civilizations across the breadth of the Old World.

Still, that's not all I mean to imply here. The technological suite that developed in the wake of the Baconian revolution will still be viable in a post-fossil fuel world, wherever the ecological and social circumstances permit it to exist at all. Deepwater shipping, canal-borne transport across nations and continents, mass production of goods using division of labor as an organizing principle, extensive use of wind and water power, and widespread literacy and information exchange involving print media, libraries, postal services, and the like, are all options that will be available to societies in the deindustrial world. So are certain other technologies that evolved in the post-Baconian era but fit neatly within the Baconian model: solar thermal technologies, for example, and those forms of electronics that can be economically manufactured and powered with the limited supplies of energy a sustainable society will have on hand.

I've suggested elsewhere that our current industrial society may turn out to be merely the first, most wasteful, and least durable of what might best be called "technic societies"—that is, human societies that get a large fraction of their total energy supply from sources other than human and animal muscle, and support complex technological suites on that basis.[2] The technologies of the Baconian era, I propose, offer a glimpse of what an emerging ecotechnic society might look like in practice—and a sense of the foundations on which the more complex ecotechnic societies of the future will build.

When the book mentioned at the beginning of this chapter claimed that "technology will always be with us," it's a safe bet that the author wasn't thinking of tall ships, canal boats, solar greenhouses, and a low-power global radio net, much less the further advances along the same lines that might well be possible in a deindustrial world. Still, it's crucial to get outside the delusion that the future must either be a flashier version of the present or a smoldering wasteland full of bleached bones, and start to confront the wider and frankly more interesting possibilities that await our descendants.

Any response to the waning of the industrial age will have to function within the constraints of a society already in the early stages of the Long Descent—an era in which energy and resources are increasingly hard for most people to obtain, in which the infrastructure that supports current lifestyles is becoming ever more brittle and prone to dysfunction, and in which most people will have to contend with the consequences of economic contraction, political turmoil, and social disintegration. As time passes, furthermore, all these pressures can be counted on to increase, and any improvement in conditions that takes place will be temporary.

All this places harsh constraints on any attempt to do anything constructive in response to the contemporary crisis of industrial civilization. Still, there are still options available that could make the decline a little less bitter, the troubled era that will follow it a little less dark, and the recovery afterwards a little easier. Compared to grand plans to save the world in a single leap, that may not sound like much—but it certainly beats sitting on one's backside daydreaming about future societies powered by green vaporware, on the one hand, or imaginary cataclysms that will relieve us of our responsibility toward the future on the other.

It's only in the imagination of true believers in the invincibility of progress that useful technologies can never be lost. As already noted, history makes the opposite case with painful clarity. Over and over again, technologies in common use during the peak years of a civilization have been lost during the dark age that followed and had to be brought in again from some other society or reinvented from scratch once the dark age was over and rebuilding could begin. It's another commonplace of history, though, that if useful technologies can be preserved during the declining years of a society, they can spread relatively rapidly through the successor states of the dark-age period and become core elements of the new civilization that follows. A relatively small number of people can preserve a technology, furthermore, by the simple acts of learning it, practicing it, and passing it on to the next generation.

Not every technology is well suited for this sort of project. The more complex a technology is, the more dependent it is on exotic materials or concentrated energy sources, and the more infrastructure it requires, the less the chance there is that it can be preserved in the face of a society in crisis. Furthermore, if the technology doesn't provide goods or services that will be useful to people during the era of decline or the dark age that follows, its chances of being preserved at all are not good at a time when resources are too scarce to divert into unproductive uses.

Seven Sustainable Technologies

Those are tight constraints. Nonetheless, I've identified seven technological suites that can be sustained on a very limited resource base, produce goods or services of value even under dark-age conditions, and could contribute mightily to the process of rebuilding if they get through the next five centuries or so.

1. Organic intensive gardening. When future historians look back on the twentieth century, I suspect that the achievement of ours that they'll consider most important is the creation of food-growing methods that build soil fertility rather than depleting it and are sustainable on a time scale of millennia. The best of the current systems of organic intensive gardening require no resource inputs other than locally available biomass, hand tools, and muscle power, and they produce a great deal of food from a relatively small piece of ground. Among the technologies included in this suite, other than the basics of soil enhancement and intensive plant and animal raising, are composting, food storage and preservation, and solar-powered season extenders such as cold frames and greenhouses.

2. Solar thermal technologies. Most of the attention given to solar energy these days focuses on turning sunlight into electricity, but electricity isn't actually that useful in terms of meeting basic human needs. Far more useful is heat, and sunlight can be used for heat with vastly greater efficiencies than it can be turned into electrical current. Water heating, space heating, cooking, food preservation, and many other useful activities can all be done by concentrating the rays of the Sun or collecting solar heat in an insulated space. Doing these things with sunlight rather than wood heat or some other fuel source will take significant stress off damaged ecosystems while meeting a great many human needs.

3. **Sustainable wood heating.** In the Earth's temperate zones, solar thermal technologies can't stand alone, and a sustainable way to produce fuel is thus high up on the list of necessities. Coppicing, a process that allows repeated harvesting of fuel wood from the same tree, and other methods of producing flammable biomass without burdening local ecosystems belong to this technological suite; so do rocket stoves and other high-efficiency means of converting wood fuel into heat.

4. **Sustainable health care.** Health care as it's practiced in the world's industrial nations is hopelessly unsustainable, dependent as it is on concentrated energy and resource inputs and planet-wide supply chains. As industrial society disintegrates, current methods of health care will have to be replaced by methods that require much less energy and other resources and can be put to use by family members and local practitioners. Plenty of work will have to go into identifying practices that belong in this suite, since the entire field is a minefield of conflicting claims issuing from the mainstream medical industry as well as alternative health care; the sooner the winnowing gets underway, the better.

5. **Letterpress printing and its related technologies.** One crucial need in an age of decline is the ability to reproduce documents from the days before things fell apart. Because the monasteries of early medieval Europe had no method of copying faster than monks with pens, much of what survived the fall of Rome was lost during the following centuries, as manuscripts rotted faster than they could be copied. In Asia, by contrast, hand-carved woodblock printing allowed documents to be mass

produced during the same era; this helps explain why learning, science, and technology recovered more rapidly in post–Tang dynasty China and post-Heian Japan than in the post-Roman West. Printing presses with movable type were made and used in the Middle Ages, and inkmaking, papermaking, and bookbinding are equally simple, so these are well within the range of craftspeople in the deindustrial dark ages ahead.

6. Low-tech shortwave radio. The ability to communicate over long distances at a speed faster than a horse can ride is another of the significant achievements of the past two centuries, and it deserves to be passed onto the future. While the scientific advances needed to work out the theory of radio required nearly three hundred years of intensive study in physics, the technology itself is simple—an ordinarily enterprising medieval European or Chinese alchemist could easily have put together a working radio transmitter and receiver, along with the metal-acid batteries needed to power them, if he had known how. The technical knowledge in the amateur radio community, which has begun to get interested in low-tech, low-power methods again after a long flirtation with high-end technologies, could become a springboard to hand-built radio technologies that could keep going after the end of industrial society.

7. Computer-free mathematics. Until recently, it didn't take a computer to crunch the numbers needed to build a bridge, navigate a ship, balance profits against losses, or do any of ten thousand other basic or not-so-basic mathematical operations. Slide rules, nomographs, tables of logarithms, or the art of double-entry bookkeeping did the job. In the future, after computers stop being economically viable to maintain and replace, those same tasks will still need to be done, but the knowledge of how to do them without a computer is at high risk of being lost. If that knowledge can be gotten back into circulation and kept viable as the computer age winds down, a great many tasks that will need to be done in the deindustrial future will be much less problematic.

It's probably necessary to note here that the reasons our descendants a few generations from now won't be using computers are economic, not technical. If you want to build and maintain computers, you need an industrial infrastructure that can manufacture integrated circuits and other electronic components, and that requires an extraordinarily complex suite of technologies, sprawling supply chains, and a vast amount of energy—all of which has to be paid for. It's unlikely that any society in the aftermath of our age will have that kind of wealth available; if any does, many other uses for that wealth would make more sense in a deindustrialized world; and in an age when human labor is again much cheaper than mechanical energy, it will be more affordable to hire people to do the routine secretarial, filing, and bookkeeping tasks currently done by computers than to find the resources to support the baroque industrial infrastructure needed to provide computers for those tasks.

The reason it's necessary to repeat this here is that whenever I point out that computers won't be economically viable in a deindustrial world, I field a flurry of outraged comments pretending that I haven't mentioned economic issues at all and insisting that computers are so cool that the future can't possibly do without them. It's as though they think a good fairy promised them something—and they aren't paying attention to all the legends about the way that fairy gifts turn into a handful of dry leaves the next morning.

Organic gardens, solar and wood heat, effective low-tech health care, printed books, shortwave radios, slide rules, and logarithms: those aren't a recipe for the kind of civilization we have today; nor are they a recipe for a kind of civilization that's existed in the past. It's precisely the inability to imagine anything besides those two alternatives that's crippled our collective ability to think about the future. One of the lessons of history is that the decline and fall of every civilization follows the same track down, but the journey back up to a new civilization almost always breaks new ground. It would be equally accurate to point out that the decline and fall of a civilization is driven by humanity in the mass, but the way back up is inevitably the work of a creative minority with its own unique take on things. The time of that minority is still far in the future, but plenty of things that can be done right now can give the creative minds of the future more options to work with.

Those of my readers who want to do something constructive about the harsh future ahead thus could do worse than to adopt one or more of the technologies I've outlined, and make a personal commitment to learning, practicing, preserving, and transmitting that technology into the future. Those who decide that some technology I haven't listed deserves the same treatment, and are willing to make an effort to get it into the waiting hands of the future, will get no argument from me. The important thing is to get off the couch and do something, because the decline is already underway and time is getting short. Fortunately, there are examples showing how the thing can be done effectively.

Captain Erikson's Equation

I have yet to hear any current futurist mention the name of Captain Gustaf Erikson of the Åland Islands and his fleet of windjammers.[3] For all I know, he's been completely forgotten now, his name and accomplishments packed away in the same dustbin of forgotten history as solar steam-engine pioneer Augustin Mouchot, his near contemporary. If so, it's high time that his footsteps sounded again on the quarterdeck of our collective imagination, because his story—and the core insight that committed him to his life-long struggle—both have plenty to teach about the realities framing the future of technology in the wake of today's era of fossil-fueled abundance.

Erikson, born in 1872, grew up in a seafaring family and went to sea as a ship's boy at the age of nine. At 19 he was the skipper of a coastal freighter working the Baltic and North Sea ports; two years later he shipped out as mate on a windjammer for deep-water runs to Chile and Australia, and eight years after that he was captain again, sailing three- and four-masted cargo ships to the far reaches of the planet. A bad fall from the rigging in 1913 left his right leg crippled, and he left the sea to become a ship owner instead, buying the first of what would become the twentieth century's last major fleet of wind-powered commercial cargo vessels.

It's too rarely remembered these days that the arrival of steam power didn't make commercial sailing vessels obsolete across the board. The ability to chug along at eight knots or so without benefit of wind was a major advantage in some contexts—naval vessels and passenger transport, for example—but coal was never cheap, and the long stretches between coaling stations on some of the world's most important trade routes meant that a significant fraction of a steamship's total tonnage had to be devoted to coal, cutting into the capacity to haul paying cargoes. For bulk cargoes over long distances, in particular, sailing ships were a good deal more economical all through the second half of the nineteenth century, and some runs remained a paying proposition for sail well into the twentieth.

That was the niche that the windjammers of the era exploited. They were huge—up to 400 feet from stem to stern—square-sided, steel-hulled ships, fitted out with more than an acre of canvas and miles of steel-wire rigging. They could be crewed by a few dozen sailors, and they hauled prodigious cargoes: up to 8,000 tons of Australian grain, Chilean nitrate—or, for that matter, coal; it was among the ironies of the age that the coaling stations that allowed steamships to refuel on long voyages were very often kept stocked by tall ships, which could do the job more economically than steamships themselves could. The markets where wind could outbid steam were lucrative enough that at the beginning of the twentieth century there were still thousands of working sailing vessels of various sizes hauling cargoes across the world's oceans.

That didn't change until bunker oil refined from petroleum ousted coal as the standard fuel for powered ships. Petroleum products carry far more energy per pound than even the best grade of coal, and the better grades of coal were beginning to run short and rise accordingly in price well before the heyday of the windjammers was over. A diesel-powered vessel refueled less often, devoted less of its tonnage to fuel, and cost much less to operate than its coal-fired equivalent. That's why Winston Churchill, as head of Britain's Admiralty, ordered the entire British Navy converted from coal to oil in the years just before the First World War and why coal-burning steamships became hard to find anywhere on the seven seas once the petroleum revolution took place. That's also why most windjammers went out of use around the same time; they could compete against coal but not against dirt-cheap diesel fuel.

Gustaf Erikson went into business as a ship owner just as that transformation was getting under way. The rush to diesel power allowed him to buy up windjammers at a fraction of their former price—his first ship, a 1,500-ton bark, cost him less than $10,000; and the pride of his fleet, the four-masted *Herzogin Cecilie*, set him back only $20,000. A tight rein on operating expenses and a careful eye on which routes were profitable kept his firm solidly in the black. The bread and butter of his business came from shipping wheat from southern Australia to Europe; Erikson's fleet and the few other windjammers still in the running would leave European ports in the northern hemisphere's autumn and sail for Spencer Gulf on Australia's southern coast, load up with thousands of tons of wheat, and then race each other home, arriving in the spring—a good skipper with a good crew could make the return trip in less than 100 days, hitting speeds upwards of fifteen knots when the winds were right.

There was money to be made that way, but Erikson's commitment to the windjammers wasn't just a matter of profit. A sentimental attachment to tall ships was arguably part of the equation, but there was another factor as well. In his later years, Erikson was fond of telling anyone who would listen that a new golden age for sailing ships was on the horizon: sooner or later, he insisted, the world's supply of coal and oil would run out, steam and diesel engines would become so many lumps of metal fit only for salvage, and those who still knew how to haul freight across the ocean with only the wind for power would have the seas, and the world's cargoes, all to themselves.

Those few books that mention Erikson at all like to portray him as the last holdout of a departed age, a man born after his time. On the contrary, he was born before his time, and lived too soon. When he died in 1947, the industrial world's first round of energy crises were still a quarter century away, and only a few lonely prophets had begun to grasp the absurdity of trying to build an enduring civilization on the ever-accelerating consumption of a finite and irreplaceable fuel supply. He had hoped that his sons would keep the windjammers running and finish the task of getting the traditions and technology of the tall ships through the age of fossil fuels and into the hands of the seafarers of the future. I'm sorry to say that that didn't happen; the profits to be made from modern freighters were too tempting, and once the old man was gone, his heirs sold off the windjammers and replaced them with diesel-powered craft.

Erikson's story is worth remembering, though, and not simply because he was an early prophet of the twilight of the industrial age. He was also one of the very first people in our age to see past the mythology of technological progress that dominated the collective imagination of his time and ours and glimpse the potentials of the strategy discussed in this book.

We can use the example that would have been dearest to his heart, the old technology of wind-powered maritime cargo transport, to explore those potentials. To begin with, it's crucial to remember that the only thing that made tall ships obsolete as a transport technology was cheap petroleum. The age of coal-powered steamships left plenty of market niches in which sailing ships were more economical than steamers. The difference, as already noted, was a matter of energy density—that's the technical term for how much energy you get out of each pound of fuel; the best grades of coal have only about half the energy density of petroleum distillates, and as you go down the scale of coal grades, energy density drops steadily. The brown coal that's commonly used for fuel these days provides, per pound, rather less than a quarter of the heat energy you get from a comparable weight of bunker oil.

As the world's petroleum reserves keep sliding down the remorseless curve of depletion, in turn, the price of bunker oil—like that of all other petroleum products—will bounce all over the map, as declining production races demand destruction down the curve. If Erikson's tall ships were still in service, it's quite possible that they would have expanded their market share during the price spikes of 2008 and 2014, though they would have been tied up in port again when the price of bunker fuel crashed thereafter. Nonetheless, as petroleum production declines, windjammers are eventually going to win out.

Yes, I'm aware that this last claim flies in the face of one of the most pervasive superstitions of our time, the faith-based insistence that whatever technology we happen to use today must always and forever be better, in every sense but a purely sentimental one, than whatever technology it replaced. The fact remains that what made diesel-powered maritime transport standard across the world's oceans was not some abstract superiority of bunker oil over wind and canvas but the simple reality that for a while, during the heyday of cheap abundant petroleum, diesel-powered freighters were more profitable to operate than any of the other options. It was always a matter of economics, and as petroleum depletion tilts the playing field the other way, the economics will change accordingly.

All else being equal, if a shipping company can make larger profits moving cargoes by sailing ships than by diesel freighters, coal-burning steamships, or some other option, the sailing ships will get the business and the other options will be left to rust in port. It really is that simple. The point at which sailing vessels become economically viable, in turn, is determined partly by the vagaries of fuel prices and partly by the cost of building and outfitting a new generation of sailing ships. Erikson's plan was to do an end run around the second half of that equation, by keeping a working fleet of windjammers in operation on niche routes until rising fuel prices made it profitable to expand into other markets. Since that didn't happen, the lag time will be significantly longer, and bunker fuel may have to price itself entirely out of certain markets—causing significant disruptions to maritime trade and to national and regional economies—before it makes economic sense to start building windjammers again.

The Economics of Sustainable Technologies

It's a source of wry amusement to me that when the prospect of sail transport gets raised, even in the greenest of peak oil circles, the immediate reaction from most people is to try to find some way to smuggle engines back onto the tall ships. Here again, though, the issue that matters is economics, not our current superstitious reverence for loud metal objects. There were plenty of ships in the nineteenth century that combined steam engines and sails in various combinations, and plenty of ships in the early twentieth century that combined diesel engines and sails the same way.

Windjammers powered by sails alone were more economical than either of these for long-range bulk transport, because engines and their fuel supplies cost money and take up tonnage that can otherwise be used for paying cargo, thus cutting substantially into profits. For that matter, it's possible that solar steam engines, or something like them, could be used as a backup power source for the windjammers of the deindustrial future. It's interesting to note that the renewable energy used for shipping in Erikson's time wasn't limited to sails; coastal freighters of the kind Erikson skippered when he was 19 were called "onkers" in Baltic Sea slang, because their windmill-powered deck pumps made a repetitive "onk-urrr, onk-urrr" noise.

Still, the same rule applies; enticing as it might be to imagine sailors on a becalmed windjammer hauling the wooden cover off a solar steam generator, expanding the folding reflector, and sending steam down below decks to drive a propeller, whether such a technology came into use would depend on whether the cost of buying and installing a solar steam engine, and the lost earning capacity due to hold space being taken up by the engine, was less than the profit to be made by getting to port a few days sooner.

Are there applications in which engines are worth having despite their drawbacks? Of course. Unless the price of biodiesel ends up at astronomical levels, or the disruptions ahead along the curve of the Long Descent cause diesel technology to be lost entirely, tugboats will probably have diesel engines for the imaginable future. So will naval vessels, since the number of major naval battles won or lost in the days of sail because the wind blew one way or another will doubtless be on the minds of many as the age of petroleum winds down. Barring a complete collapse in technology, in turn, naval vessels will no doubt still be made of steel—once cannons started firing explosive shells instead of solid shot, wooden ships became deathtraps in naval combat—but most others won't be; large-scale steel production requires ample supplies of coke, which is produced by roasting coal, and depletion of coal supplies in a post-petroleum future guarantees that steel will be much more expensive compared to other materials than it is today, or for that matter during the heyday of the windjammers.

Note that here again, the limits to technology and resource use are far more likely to be economic than technical. In purely technical terms, a maritime nation could put much of its arable land into oil crops and use that to keep its merchant marine fueled with biodiesel. In economic terms, that's a nonstarter, since the advantages to be gained by it are much smaller than the social and financial costs that would be imposed by the increase in costs for food, animal fodder, and all other agricultural products. In the same way, the technical ability to build an all-steel merchant fleet will likely still exist straight through the deindustrial future; what won't exist is the ability to do so without facing prompt bankruptcy.

That's what happens when you have to live on the product of each year's sunlight, rather than drawing down the products of half a billion years of fossil photosynthesis. Lacking fossil fuels, there are hard economic limits to how much of anything you can produce, and increasing production of one thing pretty consistently requires cutting production of something else. People in today's industrial world don't have to think like that, but their descendants in the deindustrial world will either learn how to do so or perish.

This point deserves careful study, as it's almost always missed by people trying to think their way through the technological consequences of the deindustrial future. It's not uncommon to find in online forums claims that, for example, if a society can make one gallon of biodiesel, it can make as many thousands or millions of gallons as it wants. Technically, maybe; economically, not a chance. It's as though you made $500 a week, and someone claimed you could buy as many bottles of $100-a-bottle scotch as you wanted. Au contraire, in any given week, your ability to buy expensive scotch would be limited by your need to meet other expenses such as food and rent, and some purchase plans would be out of reach even if you ignored all those other expenses and spent your entire paycheck at the liquor store. The same rule applies to societies that don't have the windfall of fossil fuels at their disposal—and once we finish burning through the fossil fuels we can afford to extract, every human society for the rest of our species' time on Earth will be described in those terms.

The one readily available way around the harsh economic impacts of fossil fuel depletion is the one that Gustaf Erikson tried but did not live to complete—the strategy of keeping an older technology in use, or bringing a defunct technology back into service, while there's still enough wealth sloshing across the decks of the industrial economy to make it easy to do so. I've suggested above that if his firm had kept the windjammers sailing, scraping out a living on whatever market niches they could find, the rising cost of bunker oil might already have made it profitable to expand into new niches; there wouldn't have been the additional challenge of finding the money to build new windjammers from the keel up, train crews to sail them, and get ships and crews through the learning curve that's inevitably a part of bringing an unfamiliar technology on line.

That same principle can be applied elsewhere. One small example is the rediscovery of the slide rule as an effective calculating device. There are still plenty of people alive today who know how to use slide rules, plenty of books that teach how to crunch numbers with a slipstick, and plenty of slide rules around. A century down the line, when slide rules will almost certainly be much more economically viable than pocket calculators, those helpful conditions might not be in place—but if people take up slide rules now for much the same reasons that Erikson kept the tall ships sailing, and make an effort to pass skills and slipsticks on to another generation, no one will have to revive or reinvent a dead technology in order to have quick accurate calculations for practical tasks such as engineering, salvage, and renewable-energy technology.

The collection of sustainable-living skills I somewhat jocularly termed "green wizardry," which I learned back in the heyday of the appropriate-tech movement in the late 1970s and early 1980s, is another case in point. Some of that knowledge, more of the attitudes that undergirded it, and nearly all the small-scale, hands-on, basement-workshop sensibility of the movement in question has vanished from our collective consciousness in the years since the Reagan-Thatcher counterrevolution foreclosed any hope of a viable future for the industrial world.

There are still enough books on appropriate tech gathering dust in used book shops, and enough in the way of living memory among those of us who were there, to make it possible to recover those things; another generation and that hope will have gone out the window. There are plenty of other possibilities along the same lines. For that matter, it's by no means unreasonable to plan on investing in technologies that may not be able to survive all the way through the decline and fall of the industrial age, if those technologies can help cushion the way down.

Whether or not it will still be possible to manufacture PV cells at the bottom of the deindustrial dark ages, to name only one example, getting them in place now on a home or local-community scale is likely to pay off handsomely when grid-based electricity becomes unreliable, as it will. The modest amounts of electricity you can expect to get from this and other renewable sources can provide critical services (for example, refrigeration and long-distance communication) that will be worth having as our current extravagant technologies unwind.

That said, all such strategies depend on having enough economic surplus on hand to get useful technologies in place before the darkness closes in. As things stand right now, as many of my readers will have had opportunity to notice already, that surplus is trickling away. Those of us who want to help make a contribution to the future along those lines had better get moving.

6

MENTATS WANTED, WILL TRAIN

THE STRATEGY OF preserving or reviving technologies for the deindustrial future now, before the accelerating curve of decline makes that task more difficult than it already is, can be applied very broadly indeed. Just now, courtesy of the final blowoff of the age of cheap energy, we have relatively easy access to plenty of information about what worked in the past. Some other resources are already becoming harder to get, but there's still time and opportunity to accomplish a great deal with what's still available.

I've already discussed some of the possibilities made possible by the strategy of deliberate technological regression, and with any luck, other people will get to work on projects of their own that I haven't even thought of. Here, though, I want to take Gustaf Erikson's logic in a direction that probably would have made the old sea dog scratch his head in puzzlement, and talk about how a certain set of mostly forgotten techniques could be put back into use right now to meet a serious unmet need in contemporary American society.

For decades now, American public life has been dominated by thoughtstoppers—short, emotionally charged declarative sentences, some of them trivial, some of them incoherent, none of them relevant, and all of them offered up as sound bites by politicians, pundits, and ordinary Americans alike, as though they meant something and proved something. The redoubtable H.L. Mencken, writing at a time when such things were not quite as universal in the American mass mind than they have become since then, called them "credos."[1]

A vast number of Americans these days gladly affirm any number of catchphrases about which they seem never to have entertained a single original thought. Those of my readers who have tried to talk about the future with their family and friends, and have strayed from the approved narrative of progress in those conversations, will be particularly familiar with the way this works. I've thought more than once of providing my readers with Bingo cards marked with the credos most commonly used to silence discussions of our future—"They'll think of something," "Technology can solve any problem," "The world's going to end soon anyway," "It's different this time," and so on—with some kind of prize for whoever fills theirs up first.

The prevalence of credos, though, is only the most visible end of the culture of acquired stupidity discussed in Chapter Two, which pervades contemporary American life and draws much of its force from the negative consequences of progress. That culture of stupidity is a major contributor to the crisis of our age, but a crisis is always an opportunity, and with that in mind, I'd like to propose that it's time for some of us, at least, to borrow a business model from the future and start preparing for future job openings as mentats.

In Frank Herbert's iconic science fiction novel *Dune*, the source of that word, a revolt against computer technology centuries before the story opened that led to a galaxy-wide ban on thinking machines—"Thou shalt not make a machine in the image of a human mind"—and a corresponding focus on developing human capacities instead of replacing them with hardware. Mentats were among the results: human beings trained from childhood to absorb, integrate, and synthesize information. Think of them as the opposite end of human potential from the sort of credo-muttering couch potatoes who make up so much of the American population these days. Ask a mentat if it really is different this time, and after he's spent thirty seconds or so reviewing the entire published literature on the subject, he'll give you a crisp first-approximation analysis explaining what's different, what's similar, which elements of each category are relevant to the situation, and what your best course of action would be in response.

Now, of course, the training programs needed to get mentats to this level of function haven't been invented yet, but the point still stands: people who know how to think, even at a less blinding pace than Herbert's fictional characters manage, are going to be far better equipped to deal with a troubled future than those who haven't. The industrial world has been conducting what amounts to a decades-long experiment to see whether computers can make human beings more intelligent, and the answer at this point is a pretty firm no. In particular, computers tend to empower decision makers without making them noticeably smarter, and the result by and large is that today's leaders are able to make bad decisions more easily and efficiently than ever before. That is to say, machines can crunch data, but it takes a mind to turn data into information, and a well-trained and well-informed mind to refine information into wisdom.

What makes a revival of the skills of thinking particularly tempting just now is that the bar is set so low. If you know how to follow an argument from its premises to its conclusion, recognize a dozen or so of the most common logical fallacies, and check the credentials of a purported fact, then you've just left most people in the United States—including the leaders of both parties and the movers and shakers of public opinion—behind you in the dust. To that basic grounding in how to think, add a good general knowledge of history and culture and a few branches of useful knowledge in which you've put some systematic study, and you're so far ahead of the pack that you might as well hang out your shingle as a mentat right away.

Now, of course, it may be a while before there's a job market for mentats—in the post-Roman world, it took several centuries for those people who preserved the considerable intellectual toolkit of the classical world to find a profitable economic niche, and that required them to deck themselves out in tall hats with moons and stars on them.[2] In the interval before the market for wizards opens up again, though, there are solid advantages to be gained by the sort of job training I've outlined, unfolding from the fact that having mental skills that go beyond muttering credos makes it possible to make accurate predictions about the future that are considerably more accurate than the ones guiding most Americans today.

This has immediate practical value in all sorts of common, everyday situations these days. When all the people you know are rushing to sink every dollar they have in the speculative swindle *du jour*, for example, you'll quickly recognize the obvious signs of a bubble in the offing, walk away, and keep your shirt while everyone else is losing theirs. When someone tries to tell you that you needn't worry about energy costs or shortages because the latest piece of energy vaporware will surely solve all our problems, you'll be prepared to ignore her and go ahead with insulating your attic, and when someone else insists that the Earth is sure to be vaporized any day now by whatever apocalypse happens to be fashionable that week, you'll be equally prepared to ignore him and go ahead with digging the new garden bed.

If these far from inconsiderable benefits tempt you, dear reader, I'd like to offer an exercise as the very first step in your mentat training. The exercise is this: the next time you catch someone (or, better yet, yourself) uttering a familiar thought-stopper about the future—"It's different this time," "They'll think of something," "There are no limits to what human beings can achieve," "The United States has an abundant supply of natural gas," or any of the other entries in the long and weary list of contemporary American credos—stop right there and *think about it*.

Is the statement true? Is it relevant? Does it address the point under discussion? Does the evidence that supports it, if any does, outweigh the evidence against it? Does it mean what the speaker thinks it means? Does it mean anything at all? These are the sorts of questions that need to be brought to bear on the unthinking assumptions so many of us apply to the future.

The Future Ain't What It Used To Be

That same sort of mental clarity is worth applying to dubious reasoning about the future even when it isn't summed up in snappy thoughtstoppers of the sort just described. I'm thinking here, among many other examples, of a project launched a few years ago by a coterie of scientists and science fiction writers who decided that what's wrong with the world today is that there are too many negative portrayals of the future in popular media. To counter this supposed deluge of unwarranted pessimism, they organized a group called Project Hieroglyph[3] and published an anthology of new, cheery, upbeat science fiction stories about marvelous new technologies that, so they insisted, could become realities within the next fifty years.

As the editor of five original science fiction anthologies about the twilight and aftermath of the industrial age, I'm hardly in a position to discourage anyone from putting together science fiction anthologies around an unpopular theme. That said, I'd ask hard questions about any claim that the anthologies I edited will somehow cause industrial civilization to decline and fall any faster than it otherwise will—and the same sort of skepticism is worth applying to Project Hieroglyph.

The contemporary crisis of industrial society isn't being caused by a lack of optimism, after all. Its roots go deep into the tough subsoil of geological and thermodynamic reality to the lethal mismatch between fantasies of endless economic growth and the hard limits of a finite planet, and to the less immediately deadly but even more pervasive mismatch between fantasies of perpetual technological progress, on the one hand, and the law of diminishing returns and the externality trap, on the other. The failure of optimism these writers are bemoaning is thus a symptom rather than a cause, and insisting that the way to solve our problems is to broadcast more optimistic notions about the future is rather like deciding that the best way to deal with flashing red warning lights on the control panel of an airplane is to put little pieces of opaque green tape over them so everything looks fine again.

It's not as though there's been a shortage of giddily optimistic visions of a gizmocentric future in recent years, after all. I grant that the most colorful works of imaginative fiction we've seen of late have come from those economists and politicians who keep insisting that the only way out of our current economic and social malaise is to do even more of the same things that got us into it. That said, any of my readers who step into a bookstore or a video store and look for something that features interstellar travel or any of the other shibboleths of the cult of progress won't have to work hard to find one.

What's happened, rather, is that such things are no longer as popular as they once were because people find that stories about bleaker futures hedged in with harsh limits are more to their taste. The question that needs to be asked, then, is why this should be the case. As I see it, there are at least three very good reasons.

First, those bleaker futures and harsh limits reflect the realities of life in contemporary America. Set aside the top twenty percent of the population by income, and Americans have on average seen their standard of living slide steadily downhill for more than four decades.[4] In 1970, as just one measure of how far things have gone, an American family with one working-class income could afford to buy a house, pay bills on time, put three square meals on the table every day, and still have enough left over for the occasional vacation or high-ticket luxury item. Now? In much of today's America, a single working-class salary isn't enough to keep a family off the streets.

That history of relentless economic decline has had a massive impact on attitudes toward the future, toward science, and toward technological progress. In 1969, it was only in the ghettos where America confined its urban poor that any significant number of people responded to the Apollo Moon landing with the sort of disgusted alienation that Gil Scott-Heron expressed so memorably in his furious ballad "Whitey on the Moon." Nowadays, a much greater number of Americans—quite possibly a majority—see the latest ballyhooed achievements of science and technology as just one more round of pointless stunts from which they won't benefit in the least.

It's easy but inaccurate to insist that they're mistaken in that assessment. Outside the narrowing circle of the well-to-do, many Americans these days spend more time coping with the problems caused by technologies than they do enjoying the benefits thereof. Most of the jobs eliminated by automation, after all, used to provide gainful employment for the poor; most of the localities that are dumping grounds for toxic waste, similarly, are inhabited by people toward the bottom of the socioeconomic pyramid, and so on down the list of unintended consequences and technological blowback. By and large, the benefits of new technology trickle up the social ladder, while the costs and burdens trickle down; this has a lot to do with the fact that the grandchildren of people who enjoyed *The Jetsons* now find *The Hunger Games* more to their taste.

Yet there's another reason for that pessimism, and it's one that the cheerleaders of the Hieroglyph Project ignore at their peril. For decades now, the great majority of the claims made about wonderful new technologies that would inevitably become part of our lives in the next few decades have turned out to be dead wrong. From jetpacks and flying cars to domed cities and vacations on the Moon, from the fission power plants that would make electricity too cheap to meter to the conquest of poverty, disease, and death itself, most of the promises offered by the propagandists and publicists of technological progress haven't happened. That has made people noticeably less impressed by further rounds of promises that likely won't come true either.

When I was a child, if I may insert a personal reflection here, one of my favorite books was *You Will Go to the Moon* by Mae and Ira Freeman. I suspect most Americans of my generation remember that book, however dimly, with its portrayal of what space travel would be like in the near future: the great conical rocket with its winged upper stage, the white doughnut-shaped space station turning in orbit, and the rest of it. I honestly expected to make that trip someday, and I was encouraged in that act of belief by a chorus of authoritative voices for whom permanent space stations, bases on the Moon, and a manned landing on Mars were a done deal by the year 2000.

Of course, in those days the United States still had a manned space program capable of putting bootprints on the Moon. We don't have one of those anymore. It's worth talking about why that is, because the same logic applies equally well to most of the other grand technological projects that were proclaimed not so long ago as the inescapable path to a shiny new future.

We don't have a manned space program anymore, to begin with, because the United States is effectively bankrupt, having committed itself in the usual manner to the sort of imperial overstretch usefully chronicled by Paul Kennedy in *The Rise and Fall of the Great Powers*[5] and cashed in its future for a temporary hegemony over most of the planet. That's the unmentionable subtext behind the disintegration of America's infrastructure and built environment, the gutting of its once-mighty industrial plant and the steady decline in standards of living already mentioned.

Britain dreamed about expansion into space when it still had an empire—the British Interplanetary Society was a major presence in space-travel advocacy in the first half of the twentieth century—and shelved those dreams when its empire went away; the United States is in the process of the same retreat. Still, there's more going on than this.

Another reason we don't have a manned space program anymore is that all those decades of rhetoric about New Worlds for Mankind never quite got around to discussing a crucial difference we've already discussed more than once: the difference between technical feasibility and economic viability. The promoters of space travel fell into the common trap of believing their own hype and convinced themselves that orbital factories, mines on the Moon, and the like would surely turn out to be paying propositions. What they forgot, of course, is what might be called the biosphere dividend: the vast array of goods and services that the Earth's natural cycles provide for human beings free of charge, which have to be paid for once you leave our home planet behind. The best current estimate for the value of that dividend, from a 1997 paper in *Nature* written by a team headed by Richard Costanza,[6] is that it's something like three times the total value of all goods and services produced by human beings.

As a very rough estimate, in other words, economic activity anywhere in the solar system other than Earth will cost around four times what it costs on Earth, even apart from transportation costs, because the services provided here for free by the biosphere have to be paid for in space or on the solar system's other worlds. That's why all the talk about space as a new economic frontier went nowhere; orbital manufacturing was tried—the Skylab program of the 1970s, the Space Shuttle, and the International Space Station in its early days, all featured attempts along those lines—and the modest advantages of freefall and ready access to hard vacuum didn't make enough of a difference to offset the costs. Thus manned space travel, like supersonic jetliners, nuclear power plants, and plenty of other erstwhile waves of the future, turned into a white elephant that could be supported only as long as massive and continuing government subsidies were forthcoming.

Those are two of the reasons why we don't have a manned space program anymore. The third is less tangible but, I suspect, the most important. It can be tracked by picking up any illustrated book about the solar system that was written before we got there, and comparing what other worlds were supposed to look like with what was actually waiting for our landers and probes.

The War against Nature

I have in front of me right now, for example, a painting of a scene on the Moon in a book published the year I was born.[7] It's a gorgeous, romantic view. Blue earthlight splashes over a crater in the foreground; further off, needle-sharp mountains catch the sunlight; the sky is full of brilliant stars. Too bad that's not what astronauts found when they got there. Nobody told the Moon it was supposed to cater to human notions of scenic grandeur, and so it presented its visitors with vistas of dull gray hillocks and empty plains beneath a flat black sky. To anybody but a selenologist, the one thing worth attention in that dreary scene was the glowing blue sphere of Earth 240,000 miles away.

For an even stronger contrast, consider the pictures beamed back by the first Viking probe from the surface of Mars in 1976, and compare that to the gaudy images of the Sun's fourth planet that were in circulation in popular culture up to that time. I remember the event tolerably well, and one of the things I remember most clearly is the pervasive sense of disappointment—of "is that all?"—shared by everyone in the room. The images from the lander didn't look like Barsoom or the arid but gorgeous setting of Ray Bradbury's *The Martian Chronicles* or any of the other visions of Mars everyone in 1970s America had tucked away in their brains. They looked instead, for all of either world, like an unusually dull corner of Nevada that had somehow been denuded of air, water, and life.

Here again, the proponents of space travel fell into the trap of believing their own hype and forgot that science fiction is no more about real futures than romance novels are about real relationships. That isn't a criticism of science fiction, by the way, though I suspect the members of Project Hieroglyph will take it as one. Science fiction is a literature of ideas, not of crass realities, and it evokes the sense of wonder that is its distinctive literary effect by dissolving the barrier between the realistic and the fantastic. What is too often forgotten, though, is that literary effects don't guarantee the validity of prophecies—in fact, they're far more likely to hide the flaws of improbable claims behind a haze of emotion.

Romance writers don't seem to have much trouble recognizing that their novels are not about the real world. Science fiction, by contrast, has suffered from an overdeveloped sense of its own importance for many years now, and too many of its authors liked to portray themselves as scouts for the onward march of humanity. (Note the presuppositions here: that humanity is going somewhere, that all of it's going in a single direction, and that this direction just happens to be defined by the literary tastes of an eccentric subcategory of twentieth-century popular fiction.) That sort of thinking led too many people in the midst of the postwar boom to forget that the universe is under no obligation to conform to our wholly anthropocentric notions of human destiny and provide us with New Worlds for Mankind just because we happen to want some.

Mutatis mutandis, that's what happened to most of the other grand visions of transformative technological progress that were proclaimed so enthusiastically over the past century or so. Most of them never happened, and those that did turned out to be far less thrilling and far more problematic than the advance billing insisted they would be. Faced with that repeated realization, a great many Americans decided—and not without reason—that more of the same gosh-wow claims were not of interest.

That's a huge shift, and it cuts to the core of some of the most basic projects and presuppositions of our age. Since 1605, when Sir Francis Bacon's *The Advancement of Learning* sketched out the first rough draft of modern scientific practice, the collection of activities we now call science has been deeply entangled with the fantasy of conquering nature. That phrase "the collection of activities we now call science" is as unavoidable here as it is awkward, because science as we now know it didn't exist at that time.

Even the word "science" had a different meaning in Bacon's era than it does today. Back then, it meant any organized body of knowledge. People in the seventeenth century could thus describe theology as "the queen of the sciences," as their ancestors had done for a thousand years, without any sense of absurdity. The word "scientist" didn't come along until the mid-nineteenth century, long after "science" had something like its modern meaning; much before then, it would have sounded as silly as "learningist" or "knowledgist," which is roughly what it would have meant, too.

To Francis Bacon, though, the knowledge and learning that counted was the kind that would enable human beings to control nature. His successors in the scientific revolution, the men who founded the Royal Society and its equivalents in other European countries, shared the same vision and rejected literary and other humanistic studies in favor of the quest for power over the nonhuman world. The crucial breakthrough—the leap to quantification—was a done deal before the Royal Society was founded in 1661; when Galileo thought of defining speed as a quantity rather than a quality, he kick started an extraordinary revolution in human thought.

Quantitative measurement, experimental testing, and public circulation of the results of research: those were the core innovations that made modern science possible. The dream of conquering nature, though, was what made modern science the focus of so large a fraction of the Western world's energies and ambitions over the past three hundred years. The role of the myth wasn't minor, or accidental. I would argue, in fact, that nothing like modern science would have emerged at all if the craving for domination over the nonhuman world hadn't caught fire in the collective imagination of the Western world.

Modern writers on the history of science have tolerably often wondered why the Greeks and Romans, equipped as they were with effective logic and a relatively complex technology, didn't have a scientific revolution of their own. The reason was actually quite simple. The Greeks and Romans, even when their own age of reason had reached its zenith of intellectual arrogance, never imagined that the rest of the universe could be made subordinate to human beings.

Believers in the traditional religions of the time saw the universe as the property of gods who delighted in punishing human arrogance; believers in the rationalist philosophies that partly supplanted those traditional religions rewrote the same concept in naturalistic terms and saw the cosmos as the enduring reality to whose laws and processes mortals had to adapt themselves or suffer. What we now think of as science was, in Greek and Roman times, a branch of philosophy, and it was practiced primarily to evoke feelings of wonder and awe at a cosmos in which human beings had their own proper and far from exalted place.

It took the emergence of a new religious sensibility, one that saw the material universe as a trap from which humanity had to extricate itself, to make the conquest of nature thinkable as a human goal. To the Christians of the Middle Ages, the world, the flesh, and the devil were the three obnoxious realities from which religion promised to save humanity. To believers in progress in the post-Christian West, the idea that the world was in some sense the enemy of the Christian believer, to be conquered by faith in Christ, easily morphed into the idea that the same world was the enemy of humanity, to be conquered in a very different sense by faith in progress empowered by science and technology.

That final step, though, required a dramatic shift in the way people in the Western world thought about the universe around them—a shift from a living world to a dead one, or to put the contrast even more precisely, from the world as an organism to the world as a machine.

The Dream of the Machine

Many people these days like to insist that of course the universe is a machine, and so is everything and everybody in it. There's a rich irony here, in that most of the people who make this claim embrace the sort of scientific-materialist atheism that rejects any suggestion that the universe has a creator or a purpose. A machine, though, is a purposive artifact; by definition, it's made by someone to accomplish something. If the universe is a machine, then it has a creator and a purpose; and if it doesn't have a creator and a purpose, logically speaking, it can't be a machine.

That sort of unintentional comedy inevitably comes into play whenever people don't think through the implications of their favorite metaphors. Still, chase that habit further along its giddy path and you'll find a deeper absurdity at work. When people say "The universe is a machine," unless they mean that statement as a poetic simile, they're engaging in a very dubious sort of logic. As Alfred Korzybski pointed out a good many years ago,[8] pretty much any time you say "This is that," unless you take the time to qualify what you mean in very careful terms indeed, you've just uttered nonsense.

What Korzybski called the "is of identity"—the use of the word "is" to represent =, the sign of equality—makes sense only in a very narrow range of uses. You can use the "is of identity" with good results in categorical definitions; when I commented above that a machine is a purposive artifact, that's what I was doing. Here is a concept, "machine"; here are two other concepts, "purposive" and "artifact"; the concept "machine" logically includes the concepts "purposive" and "artifact," so anything that can be described by the words "a machine" can also be described as "purposive" and "an artifact." That's how categorical definitions work.

Let's consider a second example, though: "A machine is a purple dinosaur." That utterance uses the same structure as the one we've just considered. I hope I don't have to prove to my readers, though, that the concept "machine" doesn't include the concepts "purple" and "dinosaur" in any but the most whimsical of senses. There are plenty of things that can be described by the label "machine," in other words, that can't be described by the labels "purple" or "dinosaur." The mere fact that some machines can be described as purple dinosaurs, by the way—electronic Barney dolls come to mind here—doesn't make the definition any less silly; it simply means that the statement "No machine is a purple dinosaur" can't be justified either.

With that in mind, let's take a closer look at the statement "The universe is a machine." As pointed out earlier, the concept "machine" implies the concepts "purposive" and "artifact," so if the universe is a machine, somebody made it for some purpose. Those of my readers who happen to belong to Christianity, Islam, or another religion that envisions the universe as the creation of one or more deities—not all religions make this claim—will find this conclusion wholly unproblematic. My atheist readers will disagree, of course, and their reaction is the one I want to discuss here. (Notice how "is" functions in the previous sentence: "the reaction of the atheists" equals "the reaction I want to discuss." This is one of the few other uses of "is" that doesn't reliably generate nonsense.)

In my experience, at least, atheists faced with the argument about the meaning of the word "machine" I've presented here pretty reliably respond with something like "It's not a machine in *that* sense." That response takes us straight to the heart of the logical problems with the "is of identity." In what sense is the universe a machine? Pursue the argument far enough, and unless the atheist storms off in a huff—which tends to happen more often than not—what you'll get amounts to "The universe and a machine share certain characteristics in common," which packs considerably less of a punch. Go further still—and at this point the atheist will almost certainly storm off in a huff—and you'll discover that the characteristics that the universe is supposed to share with a machine are all things we can't actually prove one way or another about the universe, such as whether it has a creator or a purpose.

The statement "The universe is a machine," in other words, doesn't do what it appears to do. It appears to state a categorical identity; it actually makes an unsupported generalization in absolute terms. It takes a mental model abstracted from one corner of human experience and applies it to something else. In this case, for polemic reasons, it does so in a predictably one-sided way: deductions approved by the person making the statement ("The universe is a machine; therefore it lacks life and consciousness") are acceptable, while deductions the person making the statement doesn't like ("The universe is a machine; therefore it was made by someone for some purpose") get the dismissive response noted above.

This sort of Orwellian doublethink appears all through the landscape of contemporary nonconversation and nondebate, to be sure, but the problems with the "is of identity" don't stop with its polemic abuse. Any time you say "This is that," and mean something other than "This has some features in common with that," you've just fallen victim to one of the central booby traps built into the structure of human thought.

Human beings think in categories. That's what made ancient Greek logic, which takes categories as its basic element, so massive a revolution in the history of human thinking: by watching the way that one category includes or excludes another, which is what the Greek logicians did, you can catch a very large fraction of human stupidities in the making. What Alfred Korzybski pointed out, in effect, is that there's a metalogic that the ancient Greeks didn't get to, and logical theorists since their time haven't really tackled either: the extremely murky relationship between the categories we think with and the things we experience, which don't come with category labels spray painted on them.

Here is a green plant with a woody stem. Is it a tree or a shrub? That depends on where you draw the line between those two categories, and as any botanist can tell you, that isn't an easy or an obvious thing. As long as you remember that categories exist purely within the human mind as conveniences for us to think with, you can navigate around the difficulties, but when you slip into thinking that the categories are more real than the unique and diverse things they describe, you're in deep, deep trouble.

It's not at all surprising that human thought should have such problems built into it. If you accept the Darwinian thesis that human beings evolved out of prehuman primates by the normal workings of the laws of evolution, as I do, it follows logically that our nervous systems and cognitive structures didn't evolve for the purpose of understanding truths about the cosmos. They evolved to assist us in getting food, attracting mates, fending off predators, and a range of similar, intellectually undemanding tasks. If you believe instead that human beings were created by a deity, the yawning chasm between creator and created, between an infinite and a finite intelligence, stands in the way of any claim that human beings can know the truth about the cosmos. Neither view supports the claim that a category created by the human mind is anything but a convenience to help our very modest mental powers grapple with an ultimately incomprehensible cosmos.

Any time human beings try to make sense of the universe or any part of it, in turn, they have to choose from among the available categories in an attempt to make the object of inquiry fit the capacities of their minds. That's what the founders of the scientific revolution did in the seventeenth century, by taking the category of "machine" and applying it to the universe to see how well it would fit. That was a perfectly rational choice from within their cultural and intellectual standpoint.

The founders of the scientific revolution were Christians to a man, after all, and some of them (for example, Isaac Newton) were devout even by the standards of the time. The idea that the universe had been made by someone for some purpose wasn't problematic to people who took it as given that the universe was made by God for the purpose of human salvation. It was also a useful choice in practical terms, because it allowed certain features of the universe—specifically, the behavior of masses in motion—to be accounted for and modeled with a clarity that previous categories hadn't managed to achieve.

The fact that one narrow aspect of the universe seems to behave like a machine, though, does not prove that the universe is a machine, any more than the fact that one machine happens to look like a purple dinosaur proves that all machines are purple dinosaurs. The success of mechanistic models in explaining the behavior of masses in motion proved that mechanical metaphors are good at fitting some of the observed phenomena of physics into a shape that's simple enough for human cognition to grasp, and that's all it proves. To get from that modest fact to the claim that the universe and everything in it are machines takes an intellectual leap that can't be justified by the facts. Part of the reason that leap was taken in the seventeenth century was the religious frame of scientific inquiry at that time, as already mentioned, but there was another factor, too.

It's a curious fact that mechanistic models of the universe appeared in Western European cultures, and become wildly popular there, before the machines did. In the early seventeenth century, machines played a very modest role in the life of most Europeans; most tasks were done using hand tools powered by human and animal muscle, the way they had been done since the agricultural revolution eight millennia or so before. The most complex devices available at the time were pendulum clocks, printing presses, handlooms, and the like—that is, the sort of thing that people who want to get away from technology these days like to use instead of machines!

For reasons that historians of ideas are still trying to puzzle out, though, Western European thinkers during those years were obsessed with machines and with mechanical explanations for the universe. Until Isaac Newton, theories of nature based on mechanical models didn't actually explain that much. Until the cascade of efforts to harness steam power that ended with James Watt's epochal steam engine, nearly a century after Newton, the idea that machines could elbow aside craftspeople with hand tools and animals pulling carts was an unproven hypothesis, and yet a great many people in Western Europe believed in the power of the machine as devoutly as their ancestors had believed in the power of the bones of the local saints.

A habit of thought very widespread in today's culture assumes that technological change happens first and the world of ideas changes in response to it. The facts simply won't support that claim, though. Far more often than not, the ideas come first and the technologies follow—and there's good reason why this should be so. Technologies don't invent themselves, after all. Somebody has to invent them, and then other people have to invest the resources to take them out of the laboratory and give them a role in everyday life. The decisions that drive invention and investment, in turn, are powerfully shaped by cultural forces, and these in turn are by no means as rational as the people influenced by them generally like to think.

People in Western Europe and a few of its colonies dreamed of machines and then created them. They dreamed of a universe reduced to the status of a machine, a universe made totally transparent to the human mind and totally subservient to the human will, and then set out to create it. Then they proceeded, to a frightening degree, to reshape their own minds to think solely in mechanical terms, even when doing so turned out to be hopelessly counterproductive.

An Old Kind of Science

Behind the myth of the machine, though, lies a misconception that makes the entire myth profoundly dysfunctional at this point in history. Central to the modern faith in progress is the conviction that human intelligence is the driving force behind our species' supposedly unstoppable ascent from the caves to the stars. Though that conviction appeals to our collective vanity, it leaves out more than it explains. What gave human intelligence the capacity to reshape the world in the past few centuries was the fantastic energy surplus provided by cheap and highly concentrated fossil fuels.

That's the unmentioned reality behind all that pompous drivel about humanity's dominion over nature. We figured out how to break into planetary reserves of fossil sunlight laid down over half a billion years of geological time, burned through most of it in three centuries of unthinking extravagance, and credited the resulting boom to our own supposed greatness. Lacking that treasure of concentrated energy, which humanity did nothing to create, the dream of conquering nature might never have gotten traction at all. As the modern Western world's age of reason dawned, there were other ideologies and nascent civil religions in the running to replace Christianity, and it was only the immense economic and military payoffs made possible by a fossil-fueled industrial revolution that allowed the civil religion of progress to elbow aside the competition and rise to its present dominance.[9]

As fossil fuel reserves are depleted at an ever more rapid pace, and have to be replaced by more costly and less abundant substitutes, the most basic precondition for progress is going away. These days, ongoing development in a handful of fields has to be balanced against stagnation in most others and, more crucially still, against an accelerating curve of economic decline that is making the products of science and technology increasingly inaccessible to those outside the narrowing circle of the well-to-do.

It's indicative that while the media babbles about the latest strides in space tourism for the very rich, rural counties across the United States are letting their roads revert to gravel because the price of asphalt has soared so high that the funds to pay for paving simply aren't there anymore. In that contrast, the shape of our future comes into sight. As the torrents of cheap energy that powered industrial society's heyday slow to a trickle, and the externalized costs of energy use pile up, the arrangements that once put the products of science and technology in ordinary households are coming apart.

That's not a fast process, or a straightforward one; different technologies are being affected at different rates, so that (for example) plenty of people in the United States who can't afford health care or heating fuel in the winter still have cell phones and internet access. Still, as the struggle to maintain fossil fuel production consumes a growing fraction of the industrial world's resources and capital, more and more of what used to count as a normal lifestyle in the industrial world is becoming less and less accessible to more and more people. In the process, the collective consensus that once directed prestige and funds to scientific research is slowly trickling away.

That will almost certainly mean the end of institutional science as it presently exists. It need not mean the end of science, and a weighty volume published to much fanfare and even more incomprehension a little more than a decade ago may just point to a way ahead.

I'm not sure how many of my readers were paying attention when archetypal computer geek Stephen Wolfram published his 1,264-page opus *A New Kind of Science* back in 2002. In the 1980s, Wolfram published a series of papers about the behavior of cellular automata—computer programs that produce visual patterns based on a set of very simple rules. Then the papers stopped appearing, but rumors spread through odd corners of the computer science world that he was working on some vast project along the same lines.

The rumors proved to be true; the vast project, the book just named, appeared on bookstore shelves all over the country; reviews covered the entire spectrum from rapturous praise to condemnation, though most of them also gave the distinct impression that their authors really didn't quite understand what Wolfram was talking about. Shortly thereafter, the entire affair was elbowed out of the headlines by something else, and Wolfram's book sank back out of public view—though I understand that it's still much read in those rarefied academic circles in which cellular automata are objects of high importance.

Wolfram's book, though, was not aimed at rarefied academic circles. It was trying to communicate a discovery that, so Wolfram believed, has the potential to revolutionize a great many fields of science, philosophy, and culture. Whether he was right is a complex issue—I tend to think he's on to something of huge importance, for reasons I'll explain in a bit—but it's actually less important than the method that he used to get there. With a clarity unfortunately rare in the sciences these days, he spelled out the key to his method early on in his book.

> In our everyday experience with computers, the programs that we encounter are normally set up to perform very definite tasks. But the key idea I had nearly twenty years ago—and that eventually led to the whole new kind of science in this book—was to ask what happens if one instead just looks at simple arbitrarily chosen programs, created without any specific task in mind. How do such programs typically behave?[10]

Notice the distinction here. Ordinarily, computer programs are designed to obey some human desire, whether that desire involves editing a document, sending an email, viewing pictures of people with their clothes off, snooping on people who are viewing pictures of people with their clothes off, or what have you. That's the heritage of the delusion of control discussed back in Chapter Two: like all other machines, computers are there to do what human beings tell them to do, and computer science thus tends to focus on finding ways to make computers do more things that human beings want them to do.

That same logic pervades many fields of contemporary science. The central role of experiment in scientific practice tends to foster that by directing attention away from what whole systems do when they're left alone and toward what they do when experimenters tinker with them. Too often, the result is that scientists end up studying the effects of their own manipulations to the exclusion of anything else. The alternative is to observe whole systems on their own terms—to study what they do, not in response to a controlled experimental stimulus but in response to the normal interplay between their internal dynamics and the environment around them.

That's what Wolfram did. He ran cellular automata, not to try to make them do this thing or that, but to understand the internal logic that determines what they do when left to themselves. What he discovered, to summarize well over a thousand pages of text in a brief phrase, is that cellular automata with extremely simple operating rules are capable of generating patterns as complex, richly textured, and blended of apparent order and apparent randomness, as the world of nature itself. Wolfram explains the relevance of that discovery:

> Three centuries ago science was transformed by the dramatic new idea that rules based on mathematical equations

> could be used to describe the natural world. My purpose in this book is to initiate another such transformation, and to introduce a new kind of science that is based on the much more general types of rules that can be embodied in simple computer programs.[10]

One crucial point here, to my mind, is the recognition that mathematical equations in science are simply models used to make sense of natural processes. There's been an enormous amount of confusion around that point, going all the way back to the ancient Pythagoreans, whose discoveries of the mathematical structures within musical tones, the movement of the planets, and the like led them to postulate that numbers comprised the *arche*, the enduring reality of which the changing world of our experience is but a transitory reflection.

As we've already seen, this confusion between the model and the thing modeled, between the symbol and the symbolized, is pandemic in modern thinking. Consider all the handwaving around the way that light seems to behave like a particle when subjected to one set of experiments and like a wave when put through a different set. Plenty of people who should know better treat this as a paradox, when it's nothing of the kind. Light isn't a wave or a particle, any more than the elephant investigated by the blind men in the famous story is a wall, a pillar, a rope, or what have you. "Particle" and "wave" are models derived from human sensory experience that we apply to fit our minds around some aspects of the way that light behaves, and that's all they are. They're useful, in other words, rather than true.

Thus mathematical equations provide one set of models that can be used to fit our minds around some of the ways the universe behaves. Wolfram's discovery is that another set of models can be derived from very simple rule-based processes of the kind that make cellular automata work. This additional set of models makes sense of features of the universe that mathematical models don't handle well—for example, the generation of complexity from very simple initial rules and conditions.

The effectiveness of Wolfram's models doesn't show that the universe is actually composed of cellular automata, any more than the effectiveness of mathematical models shows that the cosmos is actually made out of numbers. Rather, cellular automata and mathematical equations relate to nature the way that particles and waves relate to light: two sets of mental models that allow the brains of some far from omniscient social primates to make sense of the behavior of different aspects of a phenomenon complex enough to transcend all models.

It requires an unfashionable degree of intellectual modesty to accept that the map is not the territory, that the scientific model is merely a representation of some aspects of the reality it tries to describe. It takes even more of the same unpopular quality to back off a bit from trying to understand nature by trying to force it to jump through hoops, in the manner of too much contemporary experimentation, and turn more attention instead to the systematic observation of what whole systems do on their own terms, in their own normal environments, along the lines of Wolfram's work. Still, I'd like to suggest that both those steps are crucial to any attempt to keep science going as a living tradition in a future in which the attempt to conquer nature will have ended in nature's unconditional victory.

A huge proportion of the failures of our age, after all, unfold precisely from the inability of most modern thinkers to pay attention to what actually happens when that conflicts with blind faith in progress. It's because so much modern economic thought fixates on what people would like to believe about money and the exchange of wealth, rather than paying attention to what happens in the real world that includes these things, that predictions by economists generally amount to bad jokes at society's expense.

It's because next to nobody thinks through the implications of the laws of thermodynamics, the power laws that apply to fossil fuel deposits, and the energy cost of extracting energy from any source that so much meretricious twaddle about "limitless new energy resources" gets splashed around so freely by people who ought to know better. For that matter, the ever-popular claim that we're all going to die by some arbitrary date in the near future, and so don't have to change the way we're living now, gets its justification from a consistent refusal on the part of believers to check their prophecies of imminent doom against relevant scientific findings, on the one hand, or the last three thousand years of failed apocalyptic predictions on the other.

The sort of science that Wolfram has proposed offers one way out of that overfamiliar trap. Ironically, his "new kind of science" is in one sense a very old kind of science. Long before Sir Francis Bacon set pen to paper and began to sketch out a vision of scientific progress centered on the attempt to subject the entire universe to the human will, many of the activities we now call science were already being practiced in a range of formal and informal ways, and both of the characteristics I've highlighted above—a recognition that scientific models are simply mental approximations of nature and a focus on systematic observation of what actually happens—were far more often than not central to the way these activities were done in earlier ages.

The old Pythagoreans themselves got their mathematical knowledge by the same kind of careful attention to the way numbers behave that Wolfram applied two and a half millennia later to simple computer programs, just as Charles Darwin worked his way to the theory of evolution by patiently studying the way living things vary from generation to generation and the founders of ecology laid the foundations of a science of whole systems by systematically observing how living things behave in their own natural settings. That's very often how revolutions in scientific fundamentals get started, and whether Wolfram's particular approach is as revolutionary as he believes—I'm inclined to think that it is, though I'm not a specialist in the field—I've come to think that a general revision of science, a "Great Instauration" as Sir Francis Bacon called it, will be one of the great tasks of the age that follows ours.

The Glass Bead Game

The consequences of that general revision may go in directions that seem bizarre or even incomprehensible from the point of view of today's science. It's interesting to note that this possibility has already been explored in literature—in fact, in a major work of science fiction that once had wide popularity among the reading public. The interesting thing about this work is that, while it's set in the future and makes some very subtle speculations about that future, it hasn't been recognized as a science fiction novel at all. This was probably a good thing, because it won its author a Nobel Prize for literature, and you don't get those for science fiction. Still, it seems to me that it's past time that the work I have in mind be assigned to its proper genre. The novel is *The Glass Bead Game*, and its author was Hermann Hesse.

When I first started college, Hesse's name was one to conjure with among the young and hip. He'd developed a cult following on American campuses about the same time J.R.R. Tolkien did, and for similar reasons; though the two authors differed in just about every other way you care to think of, both wove hard questions about the presuppositions of twentieth-century industrial civilization into their fiction. Both were accordingly dismissed as unreadable by most Americans until the social changes of the late 1960s called those presuppositions into question.

When the reaction set in during the 1980s, Tolkien's work was neatly gelded by being turned into raw material for an industry of derivative fantasy that borrowed all his imagery and none of his ideas, and tacitly ignored the hard questions he posed about the lust for power welded into the heart of modern technology. Hesse's novels were harder to strip mine for cheap clichés, and so in America, at least, they were simply forgotten.

Even in the days when every other college student you met had a copy of *Siddhartha* or *Steppenwolf* tucked in a garish backpack, though, *The Glass Bead Game*—for some reason, most American editions retitled it *Magister Ludi*—was a somewhat more rarefied taste. It's a very odd story: a hagiography, more or less, compiled by a bumbling and officious scholar in the early twenty-fifth century, about a controversial figure of the previous century whose deep ambiguities of character and action go right over the narrator's head. There are plenty of things that make it a more challenging read than some of Hesse's shorter and more popular novels, but I've come to think that one of those relates directly to the theme of this book: the twenty-fourth-century setting that Hesse shows the reader in brief glimpses around the life of Magister Ludi Josephus II, a.k.a. Joseph Knecht, master of the Glass Bead Game, is not a twenty-fourth century that most people in the 1970s and early 1980s were willing to imagine.

It's one of the deft touches of the novel that Hesse paints that future with a very sparing brush, but the transition between our time and Joseph Knecht's gets explained in enough detail to make a definite kind of sense. The early twentieth century, in Hesse's future history, ushered in what later scholars would call the Age of Wars, a century-long period of prolonged and brutal violence that saw most of Europe repeatedly ravaged and the centers of global power shift decisively to other parts of the world. When lasting peace finally came, what was left of Europe tried to figure out what it was that had driven the frenzy. The answer they settled on was the profound dishonesty and political prostitution of the intellectual life of the age—a time when, to quote a professor of the Age of Wars cited by Joseph Knecht in a letter, "Not the faculty but His Excellency the General can properly determine the sum of two and two."

In the postwar era, accordingly, the scholarly professions reorganized themselves on monastic lines as ascetic orders, and each of the surviving European nations set aside a portion of land as a "pedagogic province," supported by the state but free from political interference, where talented youth could be educated, schoolteachers could be provided for the rest of the country, and scholars could pursue their research in relative security. Nearly the entire story of *The Glass Bead Game* takes place in one such region, Castalia, the pedagogic province of Switzerland. There and in equivalent provinces elsewhere, in the wake of the Age of Wars, the most gifted minds of each nation pursued research projects full time, and created a future...

If you were expecting that sentence to end "...of dramatic technological progress" or the like, think again. This is where Hesse's future history bounces right off the rails of our expectations. It's worth remembering that science fiction of the whizbang sort was widely read in the central Europe that Hesse knew. Nobody likes to talk much these days about pre–1945 central European science fiction, because a very large part of it enthusiastically pushed the aggressive authoritarian populism that got its lasting name from Mussolini's Fascist Party and helped launch the metastatic horror of Nazi Germany, but there was a lot of it, packed with the usual science fiction notions of endlessly accelerating social change driven by limitless technological advances. It's pretty clear that Hesse deliberately rejected those notions in his own work.

The future that the busy scholars of Castalia create, rather, is a period of ordinary European history differing from earlier periods mostly in its lack of war. Technology, far from progressing, stabilized after the Age of Wars, and most modern machines seem less common than in our time. A trip by railway makes one brief appearance early on. Automobiles exist, but only two of them appear in the story; one is owned by a wealthy and influential family, while the other is assigned to take a high official of the Castalian hierarchy to important meetings. Most of the time, when a character goes someplace and the mode of travel is mentioned at all, the trip is made on foot.

Other high technology isn't much more common. Broadcast media, type not specified, play a minor role in the story at one or two points, and there's some kind of projection system that allows equations to appear on a large screen as they're being written, but that's about it. Doubtless astronomers have big telescopes and the like—Castalia has astronomers, yes, but it's the only science that Hesse mentions by name. Most of the scholars of the pedagogic province work in fields such as mathematics, musicology, philology, and philosophy or take part in the jewel in Castalia's crown, the Glass Bead Game.

The Game is arguably Hesse's greatest creation, a stunningly successful piece of social science fiction so far ahead of the conventions of the genre that its implications haven't registered yet with other writers in the field. Unlike most modern thinkers, Hesse realized that historical periods value different intellectual projects; the contemporary conceit that treats technological progress as the most, or even the only, valid use of the human intellect is simply one more culturally and historically contingent judgment call, no more objectively true than the medieval belief that scholastic theology was the queen of the sciences. In twenty-fourth-century Europe attitudes have changed again, and an abstract contemplative discipline, half game and half art form, has become the defining cultural project of the time.

The Game emerged, according to Hesse's invented history, out of the fields of mathematics and musicology, as scholars found common patterns underlying the two disciplines—the structure of a geometric proof, let's say, sharing the same abstract form as a Bach fugue or a Gregorian chant. Early on, the game was played with an abacus-like device, with wires representing the conventional musical staff and glass beads of different sizes, colors, shapes, and so on—thus the name of the Game—providing a more complex alphabet in place of simple musical notes. Later on a formal mathematical script was developed; more scholarly disciplines took up the Game, finding their own abstract patterns and relating them to the musico-mathematic core; meditation exercises became part of the toolkit; public Games, attended by crowds, broadcast to large audiences and, surrounded by festivals of music and the arts, became major annual spectacles.

It's another of Hesse's defter touches that by the time of Joseph Knecht, the golden age of the Glass Bead Game is already past. Public Games that once extended for a month straight now run for two weeks at most, attended by smaller audiences and fewer public officials; the first stirrings of discontent about the funding allotted to Castalia are beginning to be heard; political events in the Far East have raised the specter of an end to the long period of European peace. How this plays out is something I'll leave to those of my readers who decide to try Hesse's novel for themselves, but it's not giving anything away to say that Hesse's sensitivity to the pace of historic change was a good deal keener than that of most other authors of science fiction.

Hesse's future Europe, it bears noting, is the kind of society that could exist and flourish in a future on the far side of the crises ahead of us. A nation or a continent in which automobiles are a rare and expensive luxury, railroads provide the bulk of what mechanized transport is needed, high technology is relatively scarce, and the values of society focus on pursuits that don't require burning up immense quantities of cheap energy, could probably get by tolerably well and provide a decent standard of living to its population, in the absence of fossil fuels. At a time when most people can't conceive of a world that lacks our current glut of cheap abundant energy without turning immediately to the fantasies of squalor and savagery that our culture habitually projects onto the inkblot patterns of the past, Hesse's novel suggests an alternative view—though he's quite clear, of course, that the route there leads through some very harsh territory.

This has implications that many of my readers will find uncomfortable. It's pretty much standard practice for every society to assume that its particular tastes and values are universal truths and to think that any society that doesn't share those tastes and values is by definition ignorant, or backward, or—well, you can fill in the putdown of your choice; there are plenty to go around. Our culture's machine fetish is a case in point. Even among those who recognize that the age of cheap energy is ending, the most common first reaction is to try to find some way to keep some favorite type of machine running—automobiles, the internet, the space program, you name it.

Among the most crucial tasks facing the pioneers of the deindustrial age, in turn, are those involved in slipping free of that now-obsolete mindset. Machines, as I think most of us have noticed by now, make very poor replacements for human beings, and the reverse is almost as true. Shifting from a machine society to a human society in the wake of the industrial age, then, is not simply a matter of replacing one set of mechanical components with another that happen to be human. It's necessary to replace attitudes, values, and expectations that are solely suited to machines—and nearly the entire modern worldview can be summed up in these terms—with the very different attitudes, values, and expectations that produce good results when applied to human beings. That, in turn, will involve transformations we can scarcely imagine today.

7

THE BUTLERIAN CARNIVAL

THE POINTS RAISED in previous chapters have implications that will not be easy to grasp. They imply, for example, that the conversations that need to happen now aren't about how to keep power flowing to the grid; they're about how to reduce our energy consumption so that we can get by without grid power, using other approaches to meet sharply reduced needs. They imply that we don't need more energy; we need much, much less, and this means in turn that we—meaning here especially the five percent of our species who live within the borders of the United States, who use so disproportionately large a fraction of the planet's energy and resources, and who produce a comparably huge fraction of the carbon dioxide that's driving global warming—need to retool our lives and our lifestyles to get by with the sort of energy consumption that most other human beings consider normal.

It doesn't require any particular genius or prescience to grasp this, merely the willingness to recognize that if something is unsustainable, sooner or later it won't be sustained. That's the unpalatable reality of the industrial era. Despite the rhetoric of universal betterment that was brandished about so enthusiastically by the propagandists of the industrial order, there were never enough of any of the necessary resources to make the extravagant lifestyles of the recent past possible for more than a small fraction of the world's population, or for more than a handful of generations.

Nearly all the members of our species who lived outside the industrial nations, and a tolerably large number who resided within them, were expected to carry most of the costs of reckless resource extraction and ecosystem disruption while receiving few if any of the benefits. They'll have plenty of company shortly: industrial civilization is winding down, but its consequences are not; and people around the world for centuries and millennia to come will have to deal with the depleted and damaged planet our actions have left them.

It's common these days, when such issues get raised, for the resulting discussions to start from the assumption that the only options are to keep existing structures running at all costs, on the one hand, or to make the leap to some untested future of further progress, on the other. There are, however, other options. In order to face them, though, it's going to be necessary to leave behind a great deal of the conventional wisdom of our time.

That's a commonplace of historical change. Every phrase on the lips of today's practical people, after all, was once a crazy notion taken seriously only by the lunatic fringe. That includes democracy, free-market capitalism, faith in progress, and all the other shibboleths of our age. In the same way, many of the principles and agendas that will shape the world of the future are likely with us already in embryonic form among scattered individuals and groups on the fringes of the collective conversation of our time. At least one of those trends, I'm convinced, has already found a home in various corners of the present day, and it's a trend that has much to say to the theme of this book.

We can start exploring that trend with the recent murmur of headlines about the e-book industry.[1] The triumphal language of an earlier decade, when e-book manufacturers and publishers insisted that the printed book was an anachronism soon destined for history's recycle bin, has quietly faded out. E-book reader sales have peaked and, in many markets, declined precipitately. They retain a market niche, especially as a venue for cheap pornography, but the bloom is off the rose. Among the young and hip, it's not hard at all to find people who got rid of their book collections in a rush of enthusiasm when e-books came out, regretted the action after it was too late, and now are slowly restocking their bookshelves while their e-book readers collect cobwebs or, at best, find use as a convenience for travel and the like.

More generally, a good many of the hottest new trends in popular culture aren't new trends at all—they're old trends revived, in many cases, by people who weren't even alive to see them the first time around. Kurt B. Reighley's lively guide *The United States of Americana* was the first and remains the best introduction to that phenomenon, which extends from burlesque shows and homebrewed bitters to backyard chickens and the revival of Victorian European martial arts. One pervasive thread that runs through the wild diversity of this emerging subculture is the simple recognition that many of these older things are better, in straightforwardly measurable senses, than their shiny modern mass-marketed not-quite-equivalents.

Within the retro subculture, as we may as well call it, a small but steadily growing number of people have taken the principle to its logical extreme and adopted the lifestyles and furnishings of an earlier decade wholesale.[2] The Chrisman family mentioned several chapters ago, with their adoption of a Victorian lifestyle, are only one example of this trend. The 1950s are another common target, and so far as I know, adopters of 1950s culture are the furthest along the process of turning into a community, but other decades are increasingly finding the same kind of welcome among those less than impressed by what today's society has on offer.

Meanwhile, the reenactment scene has expanded spectacularly in recent years from the original hearty fare of military regiments from past wars and the neo-medievalism of the Society for Creative Anachronism (SCA) to embrace almost any historical period you care to name. These aren't merely dress-up games; go to a buckskinner's rendezvous or an outdoor SCA event, for example, and you're more likely than not to see handspinners turning wool into yarn with drop spindles, a blacksmith or two laboring over a portable forge, and the like.

Other examples of the same broad phenomenon could be added to the list, but these will do for now. I'm well aware, of course, that most people have dismissed the things just named as bizarre personal eccentricities, right up there with the goldfish-swallowing and flagpole-sitting of an earlier era. I'd encourage those of my readers who had that reaction to stop, take a second look, and tease out the mental automatisms that make that dismissal so automatic a part of today's conventional wisdom. Once that's done, a third look might well be in order, because the phenomenon sketched out here marks a shift of immense importance for our future.

The World that Progress Has Made

For well over two centuries now, since it first emerged as the crackpot belief system of a handful of intellectuals on the outer fringes of their culture, the modern faith in progress has taken it as given that new things were by definition better than whatever they replaced. That assumption stands at the heart of industrial civilization's childlike trust in the irreversible march of progress toward a future among the stars. Finding ways to defend that belief even when it obviously wasn't true—when the latest, shiniest products turned out to be worse in every meaningful sense than the older products they elbowed out of the way—was among the great growth industries of the twentieth century, but there were plenty of cases in which progress really did seem to measure up to its billing. Given the steady increases of energy per capita in the world's industrial nations over the past century or so, that was a predictable outcome.

The difficulty, of course, is that the number of cases where new things really are better than what they replace has been shrinking steadily in recent decades, while the number of cases where old products are quite simply better than their current equivalents—easier to use, more effective, more comfortable, less prone to break, less burdened with unwanted side effects and awkward features, and so on—has been steadily rising. Back behind the myth of progress, like the little man behind the curtain in *The Wizard of Oz*, stand two unpalatable and usually unmentioned realities. The first is that profit, not progress, determines which products get marketed and which get round filed. The second is that making a cheaper, shoddier product and using advertising gimmicks to sell it anyway has been the standard marketing strategy across a vast range of businesses for years now.

Believers in progress used to take it for granted that sooner or later we'd get a world where everyone would live exciting, fulfilling lives brimful of miracle products and marvelous experiences. You still hear that sort of talk from the faithful now and then these days, but it's coming to sound a lot like all that talk about the glorious worker's paradise of the future sounded right around the time the Iron Curtain came down for good. In both cases, the future that was promised didn't have much in common with the one that actually showed up.

The one we got doesn't have some of the nastier features of the one the former Soviet Union and its satellites produced—well, not yet, at least—but the glorious consumer's paradise described in such lavish terms a few decades back got lost on the way to the spaceport. The future we've entered instead has turned out to be a bleak landscape of decaying infrastructure, abandoned factories, prostituted media, and steadily declining standards of living for everyone outside the narrowing circle of the affluent and privileged, with the remnants of our once-vital democratic institutions hanging above it all like rotting scarecrows silhouetted against a darkening sky.

In place of those exciting, fulfilling lives mentioned above, furthermore, we got the monotony and stress of long commutes, cubicle farms, and would-you-like-fries-with-that for the slowly shrinking fraction of our population who can find a job at all. The humor magazine *The Onion*, with its usual flair for packaging unpalatable realities in the form of deadpan humor, nailed it a while back with a faux health-news article announcing that the best thing office workers could do for their health is stand up at their desk, leave the office, and never go back.[3] Joke or not, it's not bad advice; if you're employed full time in the United States today, the average medieval peasant worked shorter hours, had more days off than you do, and kept a larger fraction of the product of their labor than you'll ever see.[4]

Then, of course, if you're like most Americans, you'll numb yourself once you get home by flopping down on the sofa and spending most of your remaining waking hours staring at little colored pictures on a glass screen. It's remarkable how many people get confused about what this action really entails. They insist that they're experiencing distant places, traveling in worlds of pure imagination, and so on through the whole litany of self-glorifying drivel the mass media likes to employ in its own praise.

Let us please be real: when you watch a program about the Amazon rain forest, you're not experiencing the Amazon rain forest; you're experiencing little colored pictures on a screen, and you're getting only as much of the experience as fits through the narrow lens of a video camera and the even narrower filter of the production process. The difference between experiencing something and watching it on TV or the internet, that is to say, is precisely the same as the difference between making love and watching pornography. In each case, the latter is a very poor substitute for the real thing.

For most people in today's America, in other words, the closest approach to the glorious consumer's paradise of the future they can ever hope to see is eight hours a day, five days a week of mindless, monotonous work under the constant pressure of management efficiency experts, if they're lucky enough to get a job at all. On top of that, they get to spend a couple of additional hours commuting and work any off-book hours the employer happens to choose to demand, in order to get a paycheck that buys a little less each month—inflation is under control, the government insists, but it's funny how prices somehow keep going up—of products that are more shoddily made, more frequently defective, and more likely to pose a serious threat to the health and well-being of their users with every passing year. Then they can go home and numb their nervous systems with those little colored pictures on the screen, showing them bland little snippets of experiences they will never have, wedged in there between the advertising.

That's the world progress has made. That's the shining future that resulted from all those centuries of scientific research and technological tinkering, all the genius and hard work and sacrifice that have gone into the project of progress. Whatever else can and can't be said about progress, one thing is clear: all things considered, even for the people who are supposed to be its beneficiaries, it's remarkably dull. Thus it's no accident that so many people are beginning to look for a better alternative.

The enthusiasm for past lifestyles and technologies described earlier, I've come to believe, is a symptom of that quest, the first stirring of wind that tell of the storm to come. People searching for a better way of living than the one our society offers these days are turning to the actual past rather than to some imaginary future. That's the immense shift mentioned earlier. What makes it even more momentous is that, by and large, it's not being done in the sort of grim puritanical spirit of humorless renunciation that today's popular culture expects from those who don't want what the consumer economy has on offer. It's being done, rather, in a spirit of celebration.

That's the thing that separates the retro future ahead of us from its partial equivalent mentioned in the previous chapter, the revolt against computers in Frank Herbert's novel *Dune* that gave rise to the mentats. There's no need for a Butlerian jihad. Once the falling cost of human labor intersects with the rising cost of energy and technology, and it becomes cheaper to hire file clerks and accountants than to maintain the gargantuan industrial machine that keeps computer technology available, computers will go away, or linger as a legacy technology for a narrowing range of special purposes until the hardware finally burns out.

If you want people to embrace a new way of looking at things, furthermore, violence, threats, and abusive language don't work, and it's even less effective to offer that new way as a ticket to virtuous misery, along the lines of the puritanical spirit noted above. That's why so much of the green propaganda of the past thirty years has done so little good. Too much of it, too often, has been pitched as a way to suffer self-righteously for the good of Gaia, and while that approach appeals to a certain number of would-be martyrs, that's not a large enough fraction of the population to matter.

The people who are ditching their Kindles and savoring books as physical objects, brewing their own beer and resurrecting other old arts and crafts, reformatting their lives in the modes of a past decade, or spending their spare time reconnecting with the customs and technologies of an earlier time—these people aren't doing any of those things out of some passion for self-denial. They're doing them because these things bring them delights that the shoddy mass-produced lifestyles of the consumer economy can't match.

The late David Fleming, in his tremendously useful sourcebook *Lean Logic*, has argued that the concept of carnival has the capacity to become a central organizing theme for the transition to the deindustrial future.[5] My sense is that he's right, and the movement away from cubicle culture and the shoddy and unsatisfying technologies of the present is a case in point. The transitions toward older technologies just discussed are movements in that direction, and they hint at potentials for transformation that have gone all but unnoticed among those who think they're facing the future ahead of us.

What these first stirrings suggest to me, in fact, is that the way forward isn't a Butlerian jihad, but a Butlerian carnival—a sensuous celebration of the world outside the cubicle farms and the glass screens. That carnival will inevitably draw most of its raw materials from eras, technologies, and customs of the past, which don't require the extravagant energy and resource inputs that the modern consumer economy demands and so will be better suited to a future defined by scarce energy and resources.

The Steampunk Future

Of course, it's one thing to point out that going back to the simpler and less energy-intensive technologies of earlier eras could help extract us from the corner into which industrial society has been busily painting itself in recent decades. It's quite another to point out that doing this can also be great fun, more so than anything that comes out of today's fashionable technologies, and in a good many cases the results include an objectively better quality of life as well.

That's not one of the canned speeches that opponents of progress are supposed to make. According to the folk mythology of modern industrial culture, since progress always makes things better, the foes of progress are assigned the role of putting on hair shirts and insisting that everyone has to suffer virtuously from a lack of progress for some reason based on sentimental superstition. The power of the collective imagination being what it is, it's not hard to find opponents of progress who say what they're expected to say. They thus fulfill their assigned role in contemporary culture, which is to stand there in their hair shirts bravely protesting until the steamroller of progress rolls right over them.

The grip of that particular bit of folk mythology on the collective imagination of our time is tight enough that when somebody brings up some other reason to oppose progress, a great many people quite literally can't absorb what's actually being said, and they respond instead to the canned speeches they expect to hear. Thus it's inevitable that whenever the idea of returning to older technologies is brought up, it's common for people to redefine that idea as a matter of sacrificing the delights of today's technology and creeping mournfully back to the unsatisfying lifestyles of an earlier day.

This is all the more ironic in that none of the people who are taking part in today's first drafts of the Butlerian carnival are saying anything of the kind. Most of them are enthusiastically talking about how much more durable, practical, repairable, enjoyable, affordable, and user–friendly older technologies are compared to the wretched plastic trash that fills the stores these days. They're discussing how much more fun it is to embrace the delights of outdated technologies than it would be to go creeping mournfully forward to the unsatisfying lifestyles of the present time. That heresy is far more than the alleged open-mindedness and intellectual diversity of our age is willing to tolerate, so it's not surprising that so many people try to pretend that nothing of the sort has been said at all.

There are subtleties to the Butlerian carnival that may not be obvious at first glance, though, and an example may help clarify this. One of the sustainable technologies that could be taken up, practiced, and passed down to the societies that will emerge out of the wreckage of ours, as mentioned in an earlier chapter, is computer-free mathematics, using slide rules and the other tools people used to crunch numbers before they handed over that chunk of their mental capacity to machines. In a discussion of that point on my blog, one of my readers—a college professor in the green-technology end of things—commented with amusement on the horrified response he'd likely get if he suggested to his students that they use a slide rule for their number-crunching activities.

Not at all, I replied; all he needed to do was stand in front of them, brandish the slide rule in front of their beady eyes, and say, "This, my friends, is a steampunk calculator."

It occurs to me that those of my readers who don't track the contemporary avant-garde may have no idea what that next to last word means. Like so many labels these days, it contains too much history to have a transparent meaning. Still, I imagine all my readers have at least heard of punk rock. During the 1980s, a mostly forgotten and largely forgettable literary movement in science fiction got labeled "cyberpunk;" the first half of the moniker referenced the way it fetishized the behavioral tics of 1980s hacker culture, and the second was given it because it made a great show, as punk rockers did, of being brash and belligerent. The phrase caught on, and during the next decade or so, every subset of science fiction that hadn't been around since Heinleins roamed the Earth got labeled fill-in-the-blankpunk by somebody or other.

Steampunk got its moniker during those years, and that's where the "-punk" came from. The "steam" is another matter. There was an alternative-history novel, *The Difference Engine* by William Gibson and Bruce Sterling, set in a world in which Victorian computer pioneer Charles Babbage launched the cybernetic revolution a century in advance with steam-powered mechanical computers. There was also a role-playing game called *Space 1889*—take a second look at those digits if you think that has anything to do with the 1970s TV show about Moonbase Alpha—that had Thomas Edison devising a means of spaceflight and putting the Victorian Earth in contact with alternate versions of Mars, Venus, and the Moon straight out of Edgar Rice Burroughs–era space fantasy.

Those and a few other sources of inspiration like them got artists, craftspeople, writers, and the like thinking about what an advanced technology might look like if the revolutions triggered by petroleum and electronics had never happened and Victorian steam-powered technology had evolved along its own course. The result is steampunk: part esthetic pose, part artistic and literary movement, part subculture, part excuse for role-playing and assorted dress-up games, and part collection of sweeping questions about some of the most basic presuppositions undergirding modern technology and the modern world.

It's very nearly an article of faith in contemporary industrial society that any advanced technology—at least until it gets so advanced that it zooms off into pure fantasy—must by definition look like ours. I'm thinking here of such otherwise impressive works of alternate history as Kim Stanley Robinson's *The Years of Rice and Salt*. Novels of this kind portray the scientific and industrial revolution happening somewhere other than Western Europe, but inevitably it's the same scientific and industrial revolution, producing the same technologies and the same social and cultural changes. This reflects the same myopia of the imagination that insists on seeing societies that don't use industrial technologies as "stuck in the Middle Ages" or "still in the Stone Age," or what have you: the insistence that all human history is a straight line of progress that leads unstoppably to us.

Steampunk challenges that on at least two fronts. First, by asking what technology would look like if the petroleum and electronics revolutions had never happened, it undercuts the common triumphalist notion that of course an advanced technology *must* look like ours, function like ours, and—ahem—support the same poorly concealed economic, political, and cultural agendas hardwired into the technology we currently happen to have. Despite such thoughtful works as John Ellis's *The Social History of the Machine Gun*, the role of such agendas in defining what counts for progress remains a taboo subject, and the idea that shifts in historical happenstance might have given rise to wholly different "advanced technologies" rarely finds its way even into the wilder ends of speculative fiction.

Outside the realm of fiction, though, it's rare to see any mention of the possibility that the technology we ended up with might not be the inevitable outcome of a scientific revolution. The boldest step in that direction I've seen so far comes from a school of historians who pointed out that the scientific revolution depended, in a very real sense, on the weather in the English Channel during a few weeks in 1688.[6] It so happened that the winds in those weeks kept the English fleet stuck in port while William of Orange carried out the last successful invasion (so far) of England by a foreign army.

As a direct result, the reign of James II gave way to that of William III, and Britain dodged the absolute monarchy, religious intolerance, and technological stasis that Louis XIV was imposing on France just then, and that most of the rest of Europe promptly copied. Because Britain took a different path—a path defined by limited monarchy, broad religious and intellectual tolerance, and the emergence of a new class of proto-industrial magnates whose wealth was not promptly siphoned off into the existing order but accumulated the masses of capital needed to build the world's first industrial economy—the scientific revolution of the late seventeenth and early eighteenth centuries was not simply a flash in the pan. Had James II remained on the throne, these historians argue, none of those things would have happened.

It shows just how thoroughly the mythology of progress has its claws buried in our imaginations that many people respond to that suggestion in an utterly predictable way—by insisting that the scientific and industrial revolutions would surely have taken place somewhere else and given rise to some close equivalent of today's technology anyway. (As previously noted, that's the underlying assumption of the Kim Stanley Robinson novel cited above and of many other works along the same lines.) At most, those who get past this notion of industrial society's Manifest Destiny imagine a world in which the industrial revolution never happened: where, say, European technology peaked around 1700 with waterwheels, windmills, square-rigged ships, and muskets, and Europe went from there to follow the same sort of historical trajectory as the Roman Empire or Tang-dynasty China.

Further extrapolations along those lines can be left to the writers of alternative history. The point being made by the writers, craftspeople, and fans of steampunk, though, cuts in a different direction. What the partly imaginary neo-Victorian tech of steampunk suggests is that another kind of advanced technology is possible: one that depends on steam and mechanics instead of petroleum and electronics, that accomplishes some of the same things our technology does by different means, and that also does different things—things that our technologies don't do, and in some cases quite possibly can't do.

Progressing toward Dystopia

It's here that steampunk levels its second and arguably more serious challenge against the ideology that sees modern industrial society as the zenith, so far, of the march of progress. While it drew its original inspiration from science fiction and role-playing games, what shaped steampunk as an esthetic and cultural movement was a sense of the difference between the elegant craftsmanship of the Victorian era and the shoddy plastic junk that fills today's supposedly more advanced culture. It's a sense that was already clear to social critics such as Theodore Roszak many decades ago. Here's Roszak's cold vision of the future awaiting industrial society, from his must-read book *Where the Wasteland Ends*:

> Glowing advertisements of undiminished progress will continue to rain down upon us from official quarters; there will always be well-researched predictions of light at the end of every tunnel. There will be dazzling forecasts of limitless affluence; there will even be much *real* affluence. But nothing will ever quite work the way the salesmen promised; the abundance will be mired in organizational confusion and bureaucratic malaise, constant environmental emergency, off-schedule policy, a chaos of crossed circuits, clogged pipelines, breakdowns in communication, overburdened social services. The data banks will become a jungle of misinformation, the computers will suffer from chronic electropsychosis. The scene will be indefinably sad and shoddy despite the veneer of orthodox optimism. It will be rather like a world's fair in its final days, when things start to sag and disintegrate behind the futuristic façades, when the rubble begins to accumulate in the corners, the chromium to grow tarnished, the neon lights to burn out, all the switches and buttons to stop working. Everything will take on that vile tackiness which only plastic can assume, the look of things decaying that were never supposed to grow old, or stop gleaming, never to cease being gay and sleek and perfect.[7]

As prophecies go, you must admit, this one was square on the mark. Roszak's nightmare vision has duly become the advanced, progressive, cutting-edge modern society in which we live today. That's what the steampunk movement is rejecting in its own way by pointing out the difference between the handcrafted gorgeousness of an older generation of technology and the "vile tackiness which only plastic can assume" that dominates contemporary products and, indeed, contemporary life.

It's an increasingly widespread recognition, and helps explain why so many people these days are into some form of reenactment. Whether it's the new Middle Ages of the Society for Creative Anachronism, the frontier culture of buckskinners and the rendezvous scene, the military-reenactment groups recreating the technologies and ambience of any number of long-ago wars, the primitive-technology enthusiasts getting together to make flint arrowheads and compete at throwing spears with atlatls, or what have you, has any other society seen so many people turn their backs on the latest modern conveniences to take pleasure in the technologies and habits of earlier times? Behind this interest in bygone technologies, I suggest, lies a concept that's even more unmentionable in polite company than the one I discussed above: the recognition that most of the time, these days, progress makes things worse.

By and large, as we all know and a few of us admit, the latest new, advanced, cutting-edge products of modern industrial society are shoddier, flimsier, and more thickly frosted with bugs, problems, and unwanted side effects than whatever they replaced. It's becoming painfully clear that we're no longer progressing toward some shiny Jetsons future, if we ever were; neither are we progressing over a cliff into a bigger and brighter apocalypse than anyone ever had before. Instead, we're progressing steadily along the downward curve of Roszak's dystopia of slow failure, into a crumbling and dilapidated world of spiraling dysfunctions hurriedly patched over, of systems that don't really work anymore but are never quite allowed to fail, in which more and more people every year find themselves shut out of a narrowing circle of paper prosperity but in which no public figure ever has the courage to mention that fact.

Set beside that bleak prospect, it's not surprising that the gritty but honest hands-on technologies and lifeways of earlier times have a significant appeal. There's also a distinct sense of security that comes from the discovery that one can actually get by, and even manage some degree of comfort, without having a gargantuan fossil-fueled technostructure on hand to meet one's every need. What intrigues me about the steampunk movement, though, is that it's gone beyond that kind of retro-tech to think about a different way in which technology could have developed—and in the process, it's thrown open the door to a reevaluation of the technologies we've got, and thus to the political, economic, and cultural agendas that the technologies we've got embody and, thus, inevitably further.

That's part of my interest, at any rate. Another part is based on the recognition that Victorian technology functioned quite effectively on a very small fraction of the energy that today's industrial societies consume. Estimates vary, but even the most industrialized countries in the world in 1860 got by on something like ten percent of the energy per capita that's thrown around in industrial nations today. The possibility therefore exists that something like a Victorian technology, or even something like the neo-Victorian extrapolations of the steampunk scene, might be viable in a future on the far side of industrial age, when the much more diffuse, intermittent, and limited energy available from renewable sources will be what we have left to work with for the rest of our species' time on this planet.

It's easy but irrelevant to point out that the mechanical and pneumatic systems of the Victorian era can't do the same things that the electronic technologies of the present can do. Of course a steampunk future wouldn't have video games, weather radar, or what have you. It would take advantage of the very different possibilities inherent in mechanical and pneumatic technology to do different things. It's only from within the tunnel vision of contemporary culture that the only conceivable kind of advanced technology is the kind that happens to produce video games and weather radars. An inhabitant of some alternate world where the petroleum and electronics revolutions never got around to happening, and something like steampunk technology became standard, could insist with equal force that a technology couldn't possibly be called advanced unless it featured funicular-morphoteny machines and photodyne nebulometers.

The difference between technological suites extends further than most people realize. It's not often remembered, for example, that paved roads of the modern type were not originally put there for automobiles. In America, and I believe in other countries as well, the first generation of what were called "macadamized" roads—the kind with a smooth surface rather than bare bricks or cobblestones—were built in response to lobbying by bicyclists. Here in the United States, the lobbying organization was the League of American Wheelmen. (There were plenty of wheelwomen as well, but the masculine gender still had collective force in the English of that time.) Their advocacy had a recreational side, but there was more to it than that.

A few people—among them the redoubtable economist Sir William Stanley Jevons—were already pointing out in the nineteenth century that exponential growth in coal consumption could not be maintained forever.[8] A great many more had begun to work out the practical implications of the soaring population of big cities in America and elsewhere, in terms of such homely but real problems as the disposal of horse manure. These concerns fed into the emergence of the bicycle as the hot new personal transport technology of the age.

Similar concerns guided the career of the brilliant French inventor Augustin Mouchot.[9] Noting that his native country had very limited coal reserves but had colonial possessions in North Africa with vast amounts of sunlight on offer, Mouchot devoted two decades of pioneering work to harnessing solar energy. His initial efforts focused on solar cookers, stills, and water pumps, and his success at these challenges encouraged him to tackle a challenge no previous inventor had managed: a solar steam engine. His first successful model was tested in 1866, and the Paris Exhibition of 1878 featured his masterpiece, a huge engine with a Sun-tracking conical reflector focusing sunlight on tubes of blackened copper; the solar engine pumped water, cooked food, distilled brandy, and ran a refrigerator. A similar model exhibited in Paris in 1880 ran a steam-driven printing press, which obligingly turned out 500 copies of *Le Journal Solaire*.

Two other technologies came out of the same era. The first commercial solar water heater hit the market in 1891 and very quickly became a common sight over much of the United States; the colder regions used them in the summertime, the Sun Belt year round, in either case with very substantial savings in energy costs. The fireless cooker or haybox was another successful and widely adopted technology of the age: a box full of insulation with a well in the center for a cooking pot, it was the slow cooker of its time, but without the electrical cord. Bring food to a boil on the stove and then pop the pot into the fireless cooker, and it finishes cooking by residual heat, again with substantial energy savings.

Such projects were on many minds in the last decades of the nineteenth century and the first decade of the twentieth. There were good reasons for that; the technology and prosperity of the Victorian era were utterly dependent on the extraction and consumption of nonrenewable resources, and for those who had eyes to see, the limits to growth were coming into sight. That's the thinking that lay behind sociologist Max Weber's eerie 1905 prediction of the future of the industrial economy: "This order is now bound to the technical and economic conditions of machine production which today determine the lives of all the individuals who are born into this mechanism, not only those directly concerned with economic acquisition, with irresistible force. Perhaps it will so determine them until the last ton of fossilized coal is burnt."[10]

It so happened that a temporary event pushed those limits back out of sight for three quarters of a century. The invention of the internal combustion engine, which turned gasoline from a waste product of lamp-fuel refining to one of the most eagerly sought products of the age, allowed the industrial societies of that time to put off the day of reckoning for a while. It wasn't just that petroleum replaced coal in many applications, though of course this happened; coal production was also propped up by an energy subsidy from petroleum—the machines that mined coal and the trains that shipped it were converted to petroleum fuel, so energy-rich petroleum could subsidize the extraction of low-grade coal reserves. If the petroleum revolution had not been an option, the twentieth century would have witnessed the sort of scenes we're seeing now: rising energy costs, externalized in various ways, and ragged but implacable economic contraction leading to decreasing energy use per capita in leading industrial nations, as an earlier and more gradual Long Descent got under way.

The appropriate-tech movement of the 1970s, which made so many promising first steps toward sustainability before it was crushed by the Reagan-Thatcher counterrevolution and the reckless drawdown of the North Slope and North Sea oil fields, has already been discussed here. What I'd like to suggest, though, is that the conservation and ecology movement of the 1970s wasn't the first attempt to face the limits of growth in modern times; it was the second. The first such attempt was in the late nineteenth century, and Augustin Mouchot, as well as the dozens of other solar and wind pioneers of that time—not to mention bicyclists!—were the first wave of sustainability pioneers, whose work deserves to be revived as much as that of the 1970s does.

Their work was made temporarily obsolete by the torrent of cheap petroleum energy that arrived around the beginning of the twentieth century. One interesting consequence of taking their existence into account is that it's easy to watch the law of diminishing returns at work in the can-kicking exercises made possible by petroleum. The first wave of petroleum energy pushed back the limits to growth for just over seventy years, from 1900 or so to 1972. The second did the same trick for around twenty-five years, from 1980 to 2005. The third—well, we're still in it, but it started in 2010 or so with the onset of the fracking bubble and isn't holding up very well just now. A few more cycles of the same kind, and the latest loudly ballyhooed new petroleum bonanza that disproves peak oil might keep the media distracted for a week.

A Road Not Yet Traveled

As a thought experiment, though, I encourage my readers to imagine what might have followed if that first great distraction never happened—if, let's say, due to some chance mutation among plankton back in the Cambrian period, carbon compounds stashed away in deepwater sediments turned into a waxy, chemically inert goo rather than into petroleum. The internal combustion engine would still have been invented, but without some immensely abundant source of liquid fuel to burn, it would have become, like the Stirling engine, an elegant curiosity useful only for a few specialized purposes. As coal reserves depleted, governments, industrial firms, and serious men of affairs doubtless would have become ever more fixated on seizing control of untapped coal mines wherever they could be found, and the twentieth century in this alternate world would likely have been ravaged by wars as destructive as the ones in our world.

At the same time, the pioneering work of Mouchot and his many peers would have become increasingly hard to ignore. Solar power was unquestionably less economical than coal, while there was coal, but as coal reserves dwindled—remember, there would be no huge diesel machines burning oceans of cheap petroleum, so no mountaintop removal mining, nor any of the other extreme coal-extraction methods so common today—pointing a conical mirror toward the Sun would rapidly become the better bet. As wars and power shifts deprived entire nations of access to what was left of the world's dwindling coal production, the same principle would have applied with even more force. Solar cookers and stills, solar pumps and engines, wind turbines and other renewable-energy technologies would have been the only viable options.

This alternate world would have had advantages that ours doesn't share. To begin with, energy use per capita in 1900 was a small fraction of current levels even in the most heavily industrialized nations, and whole categories of work currently done directly or indirectly by fossil fuels were still being done by human beings. Agriculture hadn't been mechanized, so the food supply wouldn't have been at risk; square-rigged sailing vessels were still hauling cargoes on the seas, so as the price of coal soared and steamboats stopped being economical, maritime trade and travel could readily downshift to familiar sail technology. As the new renewable-energy technologies became more widely distributed and more efficient, getting by with the energy supplied by sun and wind would have become second nature to everybody.

Perhaps, dear reader, you can imagine yourself sitting comfortably this afternoon in a café in this alternate world, about to read this book. It isn't on a glowing e-book screen; it's being serialized in a newspaper printed, as of course nearly everything is printed these days, by a solar-powered press. Before you get to the latest chapter in the serialization, you read with some interest that a Brazilian inventor has been awarded the prestigious Mouchot Prize for a solar steam engine that's far better suited to provide auxiliary power to sailing ships than existing models. You skim over the latest news from the war between Austria and Italy, in which bicycle-mounted Italian troops have broken the siege of Gemona del Friuli, and a report from Iceland, which is rapidly parlaying its abundant supply of volcanic steam into a place as one of the twenty-first century's industrial powerhouses.

It's a cool, clear, perfectly seasonable day—remember, most of the gigatons of carbon we spent the twentieth century dumping into the atmosphere stayed buried in this alternate world—and the proprietor of the café is beaming as he watches sunlight streaming through the windows. He knows that every hour of sunlight falling on the solar collectors on the roof is saving him plenty of money in expensive fuel the kitchen won't have to burn. Outside the café, the Sun gleams on a row of bicycles, yours among them: they're the normal personal transport of the twenty-first century, after all. Solar water heaters gleam on every roof, and great conical collectors track the Sun atop the factory down the road. High overhead, a dirigible soars silently past; we'll assume, for the sake of today's steampunk sensibility, that lacking the extravagant fuel supplies needed to make airplanes more than an exotic fad, the bugs got worked out of dirigible technology instead.

Back in the cafe, you begin to read the first chapter of this book—and my imagination fails me at this point, because that first chapter wouldn't have much in common with the one you've already read. A society of the sort I've just sketched out would already have learned the drawbacks of a mindset that treats linear progress as a law of nature. It would have made the transition from fossil fuels to renewable energy when its energy consumption per capita was an order of magnitude smaller than ours and, thus, would have had a much easier time of it. Of course, a more or less stable planetary climate, and an environment littered with far fewer of the ugly end products of human chemical and nuclear tinkering would be important advantages as well.

It's far from impossible that our descendants could have a society and a technology something like the one I've outlined here, though we have a long rough road to travel before that becomes possible. In the alternate world I've sketched, though, that would be no concern of mine, since ecology would be simple common sense and the unwelcome future waiting for us in this world would have gone wherever might-have-beens spend their time. That's the road we did not take—but it's also a road that could be taken, starting now, by those who choose to look past the phantom imagery of linear progress toward the far more diverse, and far more interesting, futures that can still be created by those who have the imagination to do so.

8

THE FUTURE OF CIVILIZATION

THE BUTLERIAN CARNIVAL, whatever forms it takes, is a transitional process—one of many ways of dealing with the end of progress and our civilization's rough transition into a deindustrial future. The civilizations that build on our ruins, whether or not they have any resemblance to the steampunk future I sketched out in the previous chapter, will also be transitional in nature. This is where we encounter the most challenging of all the consequences of letting go of blind faith in progress.

You might think that a society that prides itself on its accurate knowledge of the past, that devotes so much money and effort to historical monuments and museums that display the bones of prehistoric beasts, would turn something like the same attention to detail toward its future. Not so; to an astonishing degree, thinking about the future in our society tends to fall into a handful of clichés, of which the most common is ongoing progress in technology combined with near-total stasis in most other fields and a few social trends taken a bit further than anyone has yet gone.

The past has hard lessons in store for those who approach the future with such assumptions. The rise and fall of civilizations over the five thousand years of recorded history, in particular, has two very important things to say about the prospects ahead of us. The first is that there is no such thing as civilization in the abstract, only civilizations in the plural; the second is that civilizations are temporary phenomena.

Let's take a moment to unpack both of these statements. It's common, and not unreasonably so, for people today to use the word "civilization" to mean our kind of civilization, with something more or less equivalent to the technologies, values, and social habits that modern industrial society has displayed over the past three hundred years or so. Even the briefest glimpse into the past should have been enough to dispel that illusion, and it probably would have been, except for the power of the modern mythology of progress.

To the believer in progress, after all, every previous society is a way station on a road that leads straight to us, and then through us to a future that's just like the present, but even more so. That makes it easy to flatten out the profound differences that divide the civilizations of the past from each other and from us. The civilizations that will build upon our ruins, in turn, are no more likely to resemble ours than, say, Roman civilization resembles the civilization of the classic Maya. What's more, to judge by what's happened in the past, those future civilizations will pick and choose among whatever legacies from our civilization come down to them, and put them to uses that reflect their own values rather than ours.

It's the second statement, though, that tends to be most difficult for believers in progress to cope with. History shows, again, that civilizations have a limited life span; it typically takes around a thousand years for them to cycle through the normal stages of their life cycle and collapse. It's far from uncommon for a new civilization to emerge out of the ruins of the old after an interval of a few hundred years, and it's even possible for the same cycle to repeat multiple times—ancient Egypt and traditional China both managed this feat repeatedly, with relatively brief dark ages separating long intervals of civilization. It also happens, though, that the fall of one civilization is followed not by the rise of another but by a long period in which the kind of human society we call "civilization" is notable by its absence. If we're going to explore the shape of a post-progress future—the future into which we're fairly obviously headed—it's not inappropriate to include that factor in the discussion.

The Cimmerian Hypothesis

We could approach that prospect from many angles, but since Hermann Hesse and the creators of the steampunk movement have already contributed to our exploration, we may as well turn to fiction again for a guiding thread. Here's the starting place I have in mind:

> "Barbarism is the natural state of mankind," the borderer said, still staring somberly at the Cimmerian. "Civilization is unnatural. It is a whim of circumstance. And barbarism must always ultimately triumph."[1]

That's the last paragraph of "Beyond the Black River," penned by Robert E. Howard during the golden age of the weird tale in the years between the two world wars. The Cimmerian in the scene is, of course, Howard's iconic barbarian hero Conan, who has just helped the borderer and his comrades, warriors of the civilized kingdom of Aquilonia, drive back an assault by the barbarian Picts. It's easy to take the borderer's words as nothing more than a bit of bluster meant to add color to an adventure story—easy but, I'd suggest, inaccurate.

Science fiction has made much of its claim to be a "literature of ideas." A strong case can be made, though, that the weird tale, as developed by Robert E. Howard and his friends H.P. Lovecraft and Clark Ashton Smith, has at least as much claim to the same label. What's more, the ideas that feature in a classic weird tale are often a good deal more challenging than those that are the stock in trade of most science fiction: "Gee, what happens if I extrapolate this technological trend a little further?" and the like. Howard and his fellow authors, in the stories they crafted for the legendary *Weird Tales* magazine, liked to pose edgy questions about the way that the posturings of our species and its contemporary cultures appeared in the cold light of a cosmos that's wholly uninterested in our overblown opinion of ourselves.

Thus I think it's worth giving Conan and his fellow barbarians their due and treating what we may as well call the Cimmerian hypothesis as a serious proposal about the underlying structure of human history. Let's start with some basics. What is civilization? What is barbarism? What exactly does it mean to describe one state of human society as natural and another unnatural, and how does that relate to the repeated triumph of barbarism at the end of every civilization?

The word "civilization" has a galaxy of meanings, most of them irrelevant to the present purpose. We can take the original meaning of the word—in late Latin, *civilisatio*—as a workable starting point; it means "being or becoming a member of a settled community." A people known to the Romans was civilized if its members lived in *civitates*, cities or towns. We can generalize this a little further and say that a civilization is a form of human society in which people live in artificial environments. Is there more to civilization than that? Of course there is, but as I hope to show, most of it unfolds from the distinction just traced out.

A city, after all, is a human environment from which the ordinary workings of nature have been excluded, to as great an extent as the available technology permits. When you go outdoors in a city, nearly all the things you encounter have been put there by human beings. Even the trees are where they are because someone decided to put them there, not by way of the normal processes by which trees reproduce their kind and disperse their seeds. Those natural phenomena that do manage to elbow their way into an urban environment—rats, pigeons, and the like—are interlopers, and treated as such. The gradient between urban and rural settlements can be measured precisely by what fraction of the things that residents encounter is put there by human action, as compared to the fraction that was put there by ordinary natural processes.

What is barbarism? The root meaning here is a good deal less helpful. The Greek word βαρβαροι, *barbaroi*, originally meant "people who say 'bar bar bar'" instead of talking intelligibly in Greek. In Roman times that usage got bent around to mean "people outside the Empire," and thus in due time to "tribes who are too savage to speak Latin, live in cities, or give up without a fight when we decide to steal their land." Fast forward a century or two, and that definition morphed uncomfortably into "tribes who are too savage to speak Latin, live in cities, or stay peacefully on their side of the border"—enter Alaric's Visigoths, Genseric's Vandals, and the ebullient multiethnic horde that marched westward under the banners of Attila the Hun.

This is also where Conan enters the picture. In crafting his fictional Hyborian Age, which was vaguely located in time between the sinking of Atlantis and the beginning of recorded history, Howard borrowed freely from various eras of the past, but the Roman experience was an important ingredient—the story cited above drew noticeably on Roman Britain in the fourth century, though it also took elements from the Old West and elsewhere. The entire concept of a barbarian hero swaggering his way south into the lands of civilization, which Howard introduced to fantasy fiction (and which has been so freely and ineptly plagiarized since his time), has its roots in the late Roman and post-Roman experience, a time when a great many enterprising warriors did just that and when some, like Conan, became kings.

What sets barbarian societies apart from civilized ones is precisely that a much smaller fraction of the environment barbarians encounter results from human action. When you go outdoors in Cimmeria—if you're not outdoors to start with, which you probably are—nearly everything you encounter has been put there by nature. There are no towns of any size, just scattered clusters of dwellings in the midst of a mostly unaltered environment. Where your Aquilonian town dweller who steps outside may have to look hard to see anything that was put there by nature, your Cimmerian who shoulders his battle-ax and goes for a stroll may have to look hard to see anything that was put there by human beings.

What's more, there's a difference in what we might usefully call the transparency of human constructions. In Cimmeria, if you do manage to get in out of the weather, the stones and timbers of the hovel where you've taken shelter are recognizable lumps of rock and pieces of tree; your hosts smell like the pheromone-laden social primates they are; and when their barbarian generosity inspires them to serve you a feast, they send someone out to shoot a couple of deer, hack them into gobbets, and cook the result in some relatively simple manner that leaves no doubt in anyone's mind that you're all chewing on parts of dead animals. Follow Conan's route down into the cities of Aquilonia, and you're in a different world, where paint and plaster, soap and perfume, and fancy cookery, among many other things, obscure nature's contributions to the human world.

So that's our first set of distinctions. What makes human societies natural or unnatural? It's all too easy to sink into a festering swamp of unsubstantiated presuppositions here, since people in every human society think of their own ways of doing things as natural and normal, and everyone else's ways of doing the same things as unnatural and abnormal. Worse, there's the pervasive bad habit in industrial Western cultures of lumping all non-Western cultures with relatively simple technologies together as "primitive man"—as though there's only one of him, sitting there in a feathered war bonnet and a lion-skin kilt playing the didgeridoo—in order to flatten out human history into the supposedly straight line of progress that leads from the caves to the stars.

In point of anthropological fact, the notion of "primitive man" as an allegedly unspoiled child of nature is pure hokum, and very often racist hokum at that. "Primitive" cultures—that is to say, human societies that rely on relatively simple technological suites—differ from one another just as dramatically as they differ from modern Western industrial societies; and simpler technological suites don't necessarily correlate with simpler cultural forms. Traditional Australian aboriginal societies, which have extremely simple material technologies, are considered by many anthropologists to have among the most intricate cultures known anywhere, embracing stunningly elaborate systems of knowledge in which cosmology, myth, environmental knowledge, social custom, and scores of other fields normally kept separate in our society are woven together into dizzyingly complex tapestries of knowledge.

What's more, those tapestries of knowledge have changed and evolved over time. The hokum that underlies that label "primitive man" presupposes, among other things, that societies that use relatively simple technological suites have all been stuck in some kind of time warp since the Neolithic Age—think of the common habit of speech that claims that hunter-gatherer tribes are "still in the Stone Age" and so forth. Back of that habit of speech is the industrial world's irrational conviction that all human history is an inevitable march of progress that leads straight to our kind of society, technology, and so forth. That other human societies might evolve in different directions and find their own wholly valid ways of making a home in the universe is anathema to most people in the industrial world these days—even though all the evidence suggests that this way of looking at the history of human culture makes far more sense of the data than does the fantasy of inevitable linear progress toward us.

Thus traditional tribal societies are no more natural than civilizations are, in one important sense of the word "natural"; that is, tribal societies are as complex, abstract, unique, and historically contingent as civilizations are. There is, however, one kind of human society that doesn't share these characteristics—a kind of society that tends to be intellectually and culturally as well as technologically simpler than most and that recurs in astonishingly similar forms around the world and across time. This is the distinctive dark-age society that emerges in the ruins of every fallen civilization after the barbarian war leaders settle down to become petty kings, the survivors of the civilization's once-vast population get to work eking out a bare subsistence from the depleted topsoil, and most of the heritage of the wrecked past goes into history's dumpster.

If there's such a thing as a natural human society, the basic dark-age society is probably it, since it emerges when the complex, abstract, unique, and historically contingent cultures of the former civilization and its hostile neighbors have all imploded, and the survivors of the collapse have to put something together in a hurry with nothing but raw human relationships and the constraints of the natural world to guide them. Of course once things settle down, the new society begins moving off in its own complex, abstract, unique, and historically contingent direction; the dark-age societies of post-Mycenaean Greece, post-Roman Britain, post-Heian Japan and their many equivalents have massive similarities, but the new societies that emerged from those cauldrons of cultural rebirth had much less in common with one another than their dark-age forbears did.

In Howard's fictive history, the era of Conan came well before the collapse of Hyborian civilization; he was not himself a dark-age warlord, though he doubtless would have done well in that setting. The Pictish tribes whose activities on the Aquilonian frontier inspired the quotation cited earlier weren't a dark-age society, either, though if they'd actually existed, they'd have been well along the arc of transformation that turns the hostile neighbors of a declining civilization into the breeding ground of the war bands that show up to finish things off. The Picts of Howard's tale, though, were certainly barbarians—that is, they didn't speak Aquilonian, live in cities, or stay peaceably on their side of the Black River—and they were still around long after the Hyborian civilizations were gone.

That's one of the details Howard borrowed from history. By and large, human societies that don't have urban centers tend to last much longer than those that do. In particular, human societies that don't have urban centers don't tend to go through the distinctive cycle of decline and fall ending in a dark age that urbanized societies undergo so predictably. As we've seen, a core difference between civilizations and other human societies is that people in civilizations tend to cut themselves off from the immediate experience of nature to a much greater extent than the uncivilized do. Does this help explain why civilizations crash and burn so reliably, leaving the barbarians to play drinking games with mead while perched unsteadily on the ruins?

As it happens, I think it does.

The Limits of Human Intelligence

It's important to remember that human intelligence is not the sort of protean, world-transforming superpower with limitless potential that it's been labeled by the more overenthusiastic partisans of human exceptionalism. Rather, it's an interesting capacity possessed by one species of social primates, and quite possibly shared by some other animal species as well. Like every other biological capacity, it evolved through a process of adaptation to the environment—not, please note, to some abstract concept of the environment but to the specific stimuli and responses that a social primate gets from the African savanna and its inhabitants, including but not limited to other social primates of the same species. It's indicative that when our species originally spread out of Africa, it seems to have settled first in those parts of the Old World that had roughly savanna-like ecosystems, and only later worked out the bugs of living in such radically different environments as boreal forests, tropical jungles, and the like.

The interplay between the human brain and the natural environment is considerably more significant than has often been realized. For the past forty years or so, a scholarly discipline called ecopsychology has explored some of the ways that interactions with nature shape the human mind. More recently, in response to the frantic attempts of American parents to isolate their children from a galaxy of largely imaginary risks, psychologists have begun to talk about "nature deficit disorder," the set of emotional and intellectual dysfunctions that show up reliably in children who have been deprived of the normal human experience of growing up in intimate contact with the natural world.[2]

All of this should have been obvious from first principles. Studies of human and animal behavior alike have shown repeatedly that psychological health depends on receiving certain highly specific stimuli at certain stages in the maturation process. The famous experiments by Henry Harlow, who showed that monkeys raised with a mother-substitute wrapped in terrycloth grew up to be more or less normal, while those raised with a bare metal mother-substitute turned out psychotic even when all their other needs were met, are among the more famous of these; there have been many more, and many of them can be shown to affect human capacities in demonstrable ways. Children learn language, for example, only if they're exposed to speech during a certain age window. Lacking the right stimulus at the right time, the capacity to use language shuts down and apparently can't be restarted again.

In this latter example, exposure to speech is what's known as a triggering stimulus—something from outside the organism that kick starts a process that's already hardwired into the organism but will not get under way until and unless the trigger appears.

There are other kinds of stimuli that play different roles in human and animal development. The maturation of the human mind, in fact, might best be seen as a process in which inputs from the environment play a galaxy of roles, some of them of critical importance. What happens when the natural inputs that were around when human intelligence evolved get shut out and replaced by very different inputs put there by human beings?

We saw some of the answers in the second chapter of this book, and history confirms those answers in unmistakable terms. Look over the histories of fallen civilizations, and far more often than not, societies don't have to be dragged down the slope of decline and fall. Rather, they go that way at a run, convinced that the road to ruin must inevitably lead them to heaven on Earth.

The historian Arnold Toynbee wrote at length about the way that the elite classes of falling civilizations lose the capacity to come up with new responses for new situations, or even to learn from their mistakes; thus they keep trying to use the same failed policies over and over again until the whole system crashes to ruin.[3] That's an important factor, no question, but it's not just the elites who seem to lose track of the real world as civilizations go sliding down toward history's compost heap; it's the masses as well.

Those of my readers who want to see a fine example of this sort of blindness to the obvious need only check the latest headlines. Within the next decade or so, for example, the entire southern half of Florida will become unfit for human habitation due to rising sea levels, driven by our dumping of greenhouse gases into an already overloaded atmosphere.[4] Low-lying neighborhoods in Miami Beach already flood with sea water whenever a high tide and a strong onshore wind hit at the same time; one more foot of sea level rise, and salt water will pour over barriers into the remaining freshwater sources, turning southern Florida into a vast brackish swamp and forcing the evacuation of most of the millions who live there.

That's only the most dramatic of a constellation of catastrophes that are already tightening their grip on much of the United States. Out west, as I write these words, the rain forests of western Washington are burning in the wake of years of increasingly severe drought, California's vast agricultural acreage is reverting to desert, and the entire city of Las Vegas will probably be out of water—as in, you turn on the tap and nothing but dust comes out—in less than a decade. As waterfalls cascade down the seaward faces of Antarctic and Greenland glaciers, leaking methane blows craters in the Siberian permafrost, and sea level rises at rates considerably faster than the worst case scenarios scientists were considering a few years ago, these threats are hardly abstract issues. Is anyone in America taking them seriously enough to, say, take any concrete steps to stop using the atmosphere as a gaseous sewer, starting with their own personal behavior? Surely you jest.

No, the Republicans are still out there insisting at the top of their lungs that any scientific discovery that threatens their rich friends' profits must be fraudulent; the Democrats are still out there proclaiming just as loudly that there must be some way to deal with anthropogenic climate change that won't cost them their frequent-flyer miles or make them trade in their SUVs; and nearly everyone outside the political sphere is making whatever noises they think will allow them to keep on pursuing exactly those lifestyle choices that are bringing on planetary catastrophe. Every possible excuse to insist that what's already happening won't happen gets instantly pounced on as one more justification for inertia—the claim currently being splashed around the media that the Sun *might* go through a cycle of slight cooling in the decades ahead is the latest example.[5] (For the record, even if we get a grand solar minimum, its effects will be canceled out in short order by the impact of ongoing atmospheric pollution.)

Business as usual is very nearly the only option anybody is willing to discuss, even though the long-predicted climate catastrophes are already happening and the days of business as usual in any form are obviously numbered. The one alternative that gets air time is the popular fantasy of instant planetary dieoff, which gets plenty of attention because it's also an excuse for inaction. What next to nobody wants to talk about is the future that's arriving exactly as predicted: a future in which low-lying coastal regions around the country and the world have to be abandoned to the rising seas, while the Southwest and large portions of the mountain west become as inhospitable as the eastern Sahara or Arabia's Empty Quarter.

If the ice melt keeps accelerating at its present pace, we could be only a few decades from the point at which it's Manhattan Island's turn to be abandoned because everything below ground level is permanently flooded with seawater and every winter storm sends waves rolling right across the island and flings driftwood logs against second-story windows. A few decades more, and waves will roll over the low-lying neighborhoods of Houston, Boston, Seattle, and Washington DC, while the ruined buildings that used to be New Orleans rise out of the still waters of a brackish estuary and the ruined buildings that used to be Las Vegas are half buried by the drifting sand. Take a moment to consider the economic consequences of that scale of infrastructure loss and destruction of built capital, that number of people who somehow have to be evacuated and resettled, and think about what kind of body blow that will deliver to an industrial society that is already in bad shape for other reasons.

None of this had to happen. Half a century ago, policy makers and the public alike had already been presented with a tolerably clear outline of what was going to happen if we proceeded along the trajectory we were on, and those same warnings have been repeated with increasing force year by year, as the evidence to support them has mounted up implacably—and yet nearly all of us nodded and smiled and kept going. Nor has this changed in the least as the long-predicted catastrophes have begun to show up right on schedule. Quite the contrary: faced with a rising spiral of massive crises, people across the industrial world are, with majestic consistency, doing exactly those things that are guaranteed to make those crises worse.

Nature as Negative Feedback

So the question that needs to be asked, and if possible answered, is why civilizations—human societies that concentrate population, power, and wealth in urban centers—so reliably lose the capacity to learn from their mistakes and recognize that a failed policy has in fact failed. It's also worth asking why they so reliably do this within a finite and predictable timespan: as already noted, civilizations last on average around a millennium before they crash into a dark age, while uncivilized societies routinely go on for many times that period. Doubtless any number of factors drive civilizations to their messy ends, but I'd like to suggest a factor that, to my knowledge, hasn't been discussed in this context before.

Let's start with what may well seem like an irrelevancy. There's been a great deal of discussion down through the years in environmental circles about the way that the survival and health of the human body depends on inputs from nonhuman nature. There's been a much more modest amount of talk about the human psychological and emotional needs that can be met only through interaction with natural systems. One question I've never seen discussed, though, is whether the human intellect, as distinct from the body and the emotional life, has needs that are fulfilled only by a natural environment.

As I consider that question, one obvious answer comes to mind: negative feedback.

The intellect is the part of each of us that thinks, that tries to make sense of the universe of our experience. It does this by creating models. By "models" I don't just mean those tightly formalized and quantified models we call scientific theories. A poem is also a model of part of the universe of human experience; so is a myth; so is a painting; and so is a vague hunch about how something will work out. When a nine-year-old girl pulls the petals off a daisy while saying "He loves me; he loves me not," she's using a randomization technique to decide between two models of one small but, to her, very important portion of the universe, the emotional state of whatever boy she has in mind.

With any kind of model, it's critical to remember Alfred Korzybski's famous rule: "The map is not the territory."[6] A model, to put the same point another way, is a representation; as we saw back in Chapter Six, it represents the way some part of the universe looks when viewed from the perspective of one or more members of our species of social primates, using the idiosyncratic and profoundly limited set of sensory equipment, neural processes, and cognitive frameworks we got handed by our evolutionary heritage. Painful though this may be to our collective egotism, it's not unfair to say that human mental models are what you get when you take the universe and dumb it down to the point that our minds can more or less grasp it.

What keeps our models from becoming completely dysfunctional is the negative feedback we get from the universe. For the benefit of readers who didn't get introduced to systems theory, I should probably take a moment to explain negative feedback. The classic example is the common household thermostat, which senses the temperature of the air inside the house and activates a switch accordingly. If the air temperature is below a certain threshold, the thermostat turns the heat on and warms things up; if the air temperature rises above a different, slightly higher threshold, the thermostat turns the heat off and lets the house cool down.

In a sense, a thermostat embodies a very simple model of one very specific part of the universe, the temperature inside the house. Like all models, this one includes a set of implicit definitions and a set of value judgments. The definitions are the two thresholds, the one that turns the furnace on and the one that turns it off, and the value judgments label temperatures below the first threshold "too cold" and those above the second "too hot." Like every human model, the thermostat model is unabashedly anthropocentric—"too cold" by the thermostat's standard would be uncomfortably warm for a polar bear, for example—and selects out certain factors of interest to human beings from a galaxy of other things we don't happen to want to take into consideration.

The models used by the human intellect to make sense of the universe are usually more complex than the one that guides a thermostat—there are unfortunately exceptions—but they work according to the same principle. They contain definitions, which may be implicit or explicit: the girl plucking petals from the daisy may have not have an explicit definition of love in mind when she says "He loves me," but there's some set of beliefs and expectations about what those words imply underlying the model. They also contain value judgments: if she's attracted to the boy in question, "He loves me" has a positive value and "He loves me not" has a negative one.

Notice, though, that there's a further dimension to the model, which is its interaction with the observed behavior of the thing it's supposed to model. Plucking petals from a daisy, all things considered, is not a very good predictor of the emotional states of nine-year-old boys; predictions made on the basis of that method are very often disproved by other sources of evidence, which is why few girls much older than nine rely on it as an information source. Modern Western science has formalized and quantified that sort of reality testing, but it's something that most people do at least occasionally. It's when they stop doing so that we get the inability to recognize failure that helps to drive, among many other things, the fall of civilizations.

Individual facets of experienced reality thus provide negative feedback to individual models. The whole structure of experienced reality, though, is capable of providing negative feedback on another level—when it challenges the accuracy of the entire mental process of modeling.

Nature is very good at providing negative feedback of that kind. Here's a human conceptual model that draws a strict line between mammals, on the one hand, and birds and reptiles, on the other. Not much more than a century ago, it was as precise as any division in science: mammals have fur and don't lay eggs; reptiles and birds don't have fur and do lay eggs. Then some Australian settler met a platypus, which has fur and lays eggs. Scientists back in Britain flatly refused to take it seriously until some live platypuses finally made it there by ship. Plenty of platypus egg was splashed across an assortment of distinguished scientific faces, and definitions had to be changed to make room for another category of mammals and the evolutionary history necessary to explain it.

Here's another human conceptual model, the one that divides trees into distinct species. Most trees in most temperate woodlands, though, actually have a mix of genetics from closely related species. There are few red oaks; what you have instead are mostly red, partly red, and slightly red oaks. Go from the northern to the southern end of a species' distribution, or from wet to dry regions, and the variations within the species are quite often more extreme than those that separate trees that have been assigned to different species. Here's still another human conceptual model, the one that divides trees from shrubs—plenty of species can grow either way—and the list goes on.

The human mind likes straight lines, definite boundaries, precise verbal definitions. Nature doesn't. People who spend most of their time dealing with undomesticated natural phenomena, accordingly, have to get used to the fact that nature is under no obligation to make the kind of sense the human mind prefers. I'd suggest that this is why so many of the cultures that our society calls "primitive"—that is, those that have simple material technologies and interact directly with nature much of the time—so often rely on nonlogical methods of thought: those our culture labels "mythological," "magical," or "prescientific." (That the "prescientific" will inevitably turn out to be the post-scientific as well is one of the lessons of history that modern industrial society is trying its level best to ignore.) Nature as we experience it isn't simple, neat, linear, and logical, and so it makes sense that the ways of thinking best suited to dealing with nature directly aren't simple, neat, linear, and logical either.

With this in mind, let's return to the distinction discussed earlier between civilization and barbarism. A city is a human settlement from which the direct, unmediated presence of nature has been removed as completely as the available technology permits. What replaces natural phenomena in an urban setting, though, is as important as what isn't allowed there. Nearly everything that surrounds you in a city was put there deliberately by human beings; it is the product of conscious human thinking, and it follows the habits of human thought just outlined. Compare a walk down a city street to a walk through a forest or a shortgrass prairie: in the city street, much more of what you see is simple, neat, linear, and logical. A city is an environment reshaped to reflect the habits and preferences of the human mind.

I suspect there may be a straightforwardly neurological factor in all this. The human brain, so much larger compared to body weight than the brains of most of our primate relatives, evolved because having a larger brain provided some survival advantage to those hominins who had it, in competition with those who didn't. It's probably a safe assumption that processing information inputs from the natural world played a very large role in these advantages, and this would imply, in turn, that the human brain is primarily adapted for perceiving things in natural environments—not, say, for building cities, creating technologies, and making the other common products of civilization.

Thus, in order to make civilization possible, some significant part of the brain has to be redirected away from the things that it's adapted to do. I'd like to propose that the simplified, rationalized, radically information-poor environment of the city plays a crucial role in this. (Information-poor? Of course; the amount of information that comes cascading through the five keen senses of an alert hunter-gatherer standing in an African forest is vastly greater than what a city-dweller gets from the blank walls and the monotonous sounds and scents of an urban environment.) Children raised in an environment that lacks the constant cascade of information that natural environments provide, and taught to redirect their mental powers toward such other activities as reading and mathematics, grow up with cognitive habits and, in all probability, neurological arrangements focused toward the activities of civilization and away from the things to which the human brain is adapted by evolution.

One source of supporting evidence for this admittedly speculative proposal is the worldwide insistence on the part of city-dwellers that people who live in isolated rural communities, far outside the cultural ambit of urban life, are just plain stupid. What that means in practice, of course, is that people from isolated rural communities don't habitually use their brains for the particular purposes that city people value. These allegedly "stupid" country folk are, by and large, extraordinarily adept at the skills they need to survive and thrive in their own environments. They may be able to listen to the wind and know exactly where on the far side of the hill a deer waits to be shot for dinner, glance at a stream and tell which riffle the trout have chosen for a hiding place, watch the clouds pile up and read from them how many days they've got to get the hay in before the rains come and rot it in the fields—all of which tasks require sophisticated information processing, the kind of processing that human brains evolved doing.

Notice, though, how the urban environment relates to the human habit of mental modeling. Everything in a city was a mental model before it became a building, a street, an item of furniture, or what have you. Chairs look like chairs, houses like houses, and so on; it's so rare for human-made items to break out of the habitual models of our species and the particular culture that built them that when it happens, it's a source of endless comment. Where a natural environment constantly challenges human conceptual models, an urban environment reinforces them, producing a feedback loop that's probably responsible for most of the achievements of civilization.

I suggest, though, that the same feedback loop may also play a very large role in the self-destruction of civilizations. People raised in urban environments come to treat their mental models as realities, more real than the often-unruly facts on the ground, because everything they encounter in their immediate environments reinforces those models. As the models become more elaborate and the cities become more completely insulated from the complexities of nature, the inhabitants of a civilization move deeper and deeper into a landscape of hallucinations—not least because as many of those hallucinations get built in brick, stone, glass, and steel as the available technology permits. As a civilization approaches its end, the divergence between the world as it exists and the mental models that define the world for the civilization's inmates becomes total, and its decisions and actions become lethally detached from reality.

Civilization as Positive Feedback

It's from this standpoint that we can begin to understand why civilizations crash and burn so reliably. There are plenty of factors pushing in this direction, and it's most likely that several of them are responsible. The collapse of civilization could be an overdetermined process: like the victim in the cheap mystery novel who was shot, stabbed, strangled, clubbed over the head, and then chucked out a twentieth floor window, civilizations that fall may have more causes of death than were actually necessary.

The ecological costs of building and maintaining cities, for example, place much greater strains on the local environment than do the less costly and concentrated settlement patterns of non-urban societies, and the rising maintenance costs of capital—the driving force behind the theory of catabolic collapse I've proposed elsewhere[7]—can spin out of control much more easily in an urban setting than elsewhere. Other examples of the vulnerability of urbanized societies can easily be worked out by those who wish to do so. That said, the perspectives discussed in this chapter imply that there's at least one other factor at work.

People who live in a mostly natural environment—and by this I mean merely an environment in which most things are put there by nonhuman processes rather than by human action—have to deal constantly with the inevitable mismatches between the mental models of the universe they carry in their heads and the universe that actually surrounds them. People who live in a mostly artificial environment—an environment in which most things were made and arranged by human action—don't have to deal with this anything like so often, because an artificial environment embodies the ideas of the people who constructed and arranged it. A natural environment therefore applies negative or, as it's also called, corrective feedback to human models of the way things are, while an artificial environment applies positive feedback—the sort of thing people usually mean when they talk about a feedback loop.

This explains, incidentally, one of the other common differences between civilizations and other kinds of human society: the pace of change. Anthropologists not so long ago used to insist that "primitive societies" were stuck in some kind of changeless stasis. That's nonsense, as already noted, but the thin basis in fact that was used to justify the nonsense was simply that the pace of change in low-tech, non-urban societies, when they're left to their own devices, tends to be fairly sedate and usually happens over a time scale of generations. Urban societies, on the other hand, change quickly, and the pace of change tends to accelerate over time: a dead giveaway that a positive feedback loop is at work.

Notice that what's fed back to the minds of civilized people by their artificial environment isn't simply human thinking in general. It's whatever particular set of mental models and habits of thought happen to be most popular in their civilization. Modern industrial civilization, as discussed back in Chapter Two, has a fixation on machines that pervades its mental world as thoroughly as machines themselves pervade its built environment. As we've already seen, this fixation has had disastrous effects on the capacity of our civilization's elites to grapple with less mechanical realities—and the same dysfunctional thinking trickles down all the way through our society.

Another core element of the modern industrial mindset is an obsession with simplicity. Our mental models and habits of thought value straight lines, simple geometrical shapes, hard boundaries, and clear distinctions. That obsession, and the models and mental habits that unfold from it, have given us an urban environment full of straight lines, simple geometrical shapes, hard boundaries, and clear distinctions—and thus reinforce our unthinking assumption that these things are normal and natural, which by and large they aren't.

Modern industrial civilization is also obsessed with the frankly weird belief that growth for its own sake is a good thing. (Outside of a few specific cases, that is. I've wondered at times whether the deeply neurotic North American attitude toward body weight comes from the conflict between current fashions in body shape and the growth-is-good mania of the rest of our culture; if bigger is better, why isn't a big belly better than a small one?) In a modern urban North American environment, it's easy to believe that growth is good, since that claim is endlessly rehashed whenever some new megawhatsit replaces something of merely human scale and since so many of the costs of malignant growth get hauled out of sight and dumped on somebody else. In settlement patterns that haven't been pounded into their present shape by true believers in industrial society's growth-for-its-own-sake ideology, people are rather more likely to grasp the meaning of the words "too much."

I've used examples from our own civilization because they're familiar, but every civilization reshapes its urban environment in the shape of its own mental models, which then reinforce those models in the minds of the people who live in that environment. As these people in turn shape that environment, the result is positive feedback: the mental models in question become more and more deeply entrenched in the built environment and thus also the collective conversation of the culture, and in both cases they also become more elaborate and more extreme. The history of architecture in the Western world over the past few centuries is a great example of this: over that time, buildings became ever more completely defined by straight lines, flat surfaces, simple geometries, and hard boundaries between one space and another—and it's hardly an accident that so many dimensions of popular culture in urban communities have simplified in much the same way over that same timespan.

One way to understand this is to see a civilization as the working out in detail of some specific set of ideas about the world.[8] At first those ideas are as inchoate as dream images, barely grasped even by the keenest thinkers of the time. Gradually, though, the ideas get worked out explicitly; conflicts among them are resolved or papered over in standardized ways; the original set of ideas becomes the core of a vast, ramifying architecture of thought that defines the universe to the inhabitants of that civilization. Eventually, everything in the world of human experience is assigned some place in that architecture of thought; everything that can be hammered into harmony with the core set of ideas has its place in the system, while everything that can't gets assigned the status of superstitious nonsense or whatever other label the civilization likes to use for the realities it denies.

The further the civilization develops, though, the less it questions the validity of the basic ideas themselves, and the urban environment is a critical factor in making this happen. By limiting, as far as possible, the experiences available to influential members of society to those that fit the established architecture of thought, urban living makes it much easier to confuse mental models with the universe those models claim to describe, and that confusion is essential if enough effort, enthusiasm, and passion are to be directed toward the process of elaborating those models to their furthest possible extent.

A branch of knowledge that has to keep on going back to revisit its first principles, after all, will never get far beyond them. This is why philosophy, which is the science of first principles, doesn't "progress" in the simple-minded sense of that word—Aristotle didn't disprove Plato, nor did Nietzsche refute Schopenhauer, because each of these philosophers, like all others in that challenging field, returned to the realm of first principles from a different starting point and so offered a different account of the landscape. Original philosophical inquiry thus plays a very large role in the intellectual life of every civilization early in the process of urbanization, since this helps elaborate the core ideas on which the civilization builds its vision of reality. Once that process is more or less complete, though, philosophy turns into a recherché intellectual specialty or gets transformed into intellectual dogma.

Cities are thus the Petri dishes in which civilizations ripen their ideas to maturity—and like Petri dishes, they do this by excluding contaminating influences. It's easy, from the perspective of a falling civilization like ours, to see this as a dreadful mistake, a withdrawal from contact with the real world in order to pursue an abstract vision of things increasingly detached from everything else. That's certainly one way to look at the matter, but there's another side to it as well.

Civilizations are far and away the most spectacularly creative form of human society. Over the course of its thousand-year lifespan, the inhabitants of a civilization will create many orders of magnitude more of the products of culture—philosophical and religious traditions, works of art and the traditions that produce and sustain them, and so on—than an equal number of people living in non-urban societies and experiencing the very sedate pace of cultural change already mentioned. To borrow a metaphor from the plant world, non-urban societies are perennials and civilizations are showy annuals that throw all their energy into the flowering process. Having flowered, civilizations then go to seed and die, while the perennial societies flower less spectacularly and remain green thereafter.

The feedback loop described above explains both the explosive creativity of civilizations and their equally explosive downfall. It's precisely because civilizations free themselves from the corrective feedback of nature, and divert an ever larger portion of their inhabitants' brainpower from the uses for which human brains were originally adapted by evolution, that they generate such torrents of creativity. Equally, it's precisely because they do these things that civilizations run off the rails into self-feeding delusion, lose the capacity to learn the lessons of failure or even notice that failure is taking place, and are destroyed by threats they've lost the capacity to notice, let alone overcome. Meanwhile, other kinds of human societies move sedately along their own life cycles, and their creativity and their craziness—and they have both of these, of course, just as civilizations do—are kept within bounds by the enduring negative feedback loops of nature.

Which of these two options is better? That's a question of value, not of fact, and so it has no one answer. Facts belong to the senses and the intellect, and they're objective, at least to the extent that others can say, "Yes, I see it too." Values, by contrast, are a matter of the heart and the will, and they're subjective; to call something good or bad doesn't state an objective fact about the thing being discussed. It always expresses a value judgment from some individual point of view. You can't say "X is better than Y," and mean anything by it, unless you're willing to field such questions as "Better by what criteria?" and "Better for whom?"

Myself, I'm very fond of the benefits of civilization. I like hot running water, public libraries, the rule of law, and a great many other things that you get in civilizations and generally don't get outside of them. Of course that preference is profoundly shaped by the fact that I grew up in a civilization; if I'd happened to be the son of yak herders in central Asia or tribal horticulturalists in upland Papua New Guinea, I might well have a different opinion—and I might also have a different opinion even if I'd grown up in this civilization but had different needs and predilections. Robert E. Howard was a child of American civilization at its early twentieth-century zenith, and he loathed civilization and all it stood for.

This is one of the two reasons that I think it's a waste of time to get into arguments over whether civilization is a good thing. The other reason is that neither my opinion nor yours, dear reader, nor the opinion of anybody else who might happen to want to fulminate on the internet about the virtues or vices of civilization, is going to have any effect on whether civilizations exist in the future. Since the basic requirements of urban life first became available not long after the end of the last ice age, civilizations have risen wherever conditions favored them, cycled through their lifespans, and fallen, and new civilizations have risen again in the same places if the conditions remained favorable for that process. That will likely continue as far into the future as our species endures.

Thus I'm not proposing, on the basis of the argument just set out, that we all ought to abandon cities once and for all and return to modes of living that aren't quite so good at screening out the negative feedback that nature provides our intellects. In the post-progress future, some people will do that, by choice or from necessity. There are already people moving in that direction, dropping out of today's industrial lifestyles and adopting ways of life that make more sense to them, and I expect more to follow the same path as the benefits of industrial lifestyles decline and the costs pile up. It's a valid choice.

At the same time, trying to preserve as many of the benefits of civilization as possible, for as long as possible, and pass them onto the builders of future urban societies is also a valid choice. There are many roads that lead into the post-progress future.

A Camp amid the Ruins

Until the coming of the fossil fuel age, civilization was a localized thing, in a double sense. On the one hand, without the revolution in transport and military technology made possible by fossil fuels, any given civilization could maintain control over only a small portion of the planet's surface for more than a fairly short time—thus as late as 1800, when the industrial revolution was already well under way, the civilized world was still divided into separate civilizations that each pursued its own very different ideas and values.

On the other hand, without the economic revolution made possible by fossil fuels, very large sections of the world were unsuited to civilized life and remained outside the civilized world for all practical purposes. As late as 1800, as a result, quite a bit of the world's land surface was still inhabited by hunter-gatherers, nomadic pastoralists, and tribal horticulturalists who owed no allegiance to any urban power and had no interest in cities at all—except for the nomadic pastoralists, that is, who occasionally liked to pillage one.

For the rest of the time our species survives, we will have much more sharply constrained energy supplies than we've had handy over the past few centuries. That doesn't mean, as already noted, that our descendants will be condemned to huddle in caves until the jaws of extinction close around them. Rather, it means that the limits to civilization that existed before 1800 will most likely return, and as civilizations begin to build on our ruins, using something like the suites of sustainable technologies already discussed in this book, they will be much more limited, localized, and diverse than the industrial societies of today.

I've argued at some length in an earlier book[9] that, many centuries from now, the most likely endpoint of the mess we're currently in will be the emergence of at least a few ecotechnic societies—societies that can create and maintain relatively high technology on the modest energy and resource inputs that can be provided by renewable sources. I've also suggested, there and elsewhere, that there's quite a bit that can be done here and now to lay the foundations for the ecotechnic societies of the far future, and that there's quite a bit that can be done here and now to make the unraveling of the age of abundance less traumatic than it will otherwise be.

To my mind, those are worthwhile goals. What makes them difficult is simply that any meaningful attempt to pursue them has to start by accepting that the age of cheap, abundant energy is ending, that the lifestyles made possible by that age are ending with it, and that wasting all those fossil fuels on what amounts to a drunken binge three centuries long might not have been a very smart idea in the first place. Any one of those would be a bitter pill to take; all three of them together are far more than most people nowadays are willing to swallow, and so it's not surprising that so much effort over the past few decades has gone into pretending that the squalid excesses of contemporary culture can somehow keep rolling along in the teeth of all the evidence to the contrary.

The world's fossil fuel reserves, and its other nonrenewable resources, won't come back on any time scale that matters to human beings. Since we've burned all the easily accessible coal, oil, and natural gas on the planet, and are working our way through the stuff that's difficult to get even with today's baroque and energy-intensive technologies, the world's first fossil-fueled human civilization is guaranteed to be its last as well. Thus the shape of the future will have much more in common with the shape of the past than the modern mythology of progress allows.

As new civilizations arise, the energy resources they'll have available to them will be far less abundant and concentrated than the fossil fuels that gave industrial civilization its global reach. With luck, and some hard work on the part of people living now, they may well inherit the information they need to make use of sun, wind, and other renewable energy resources in ways that the civilizations before ours didn't know how to do. As our present-day proponents of green energy are finding out the hard way just now, though, this doesn't amount to the kind of energy necessary to maintain our kind of civilization.

It may be, as I've suggested, that modern industrial society is simply the first, clumsiest, and most wasteful form of technic society, the subset of human societies that get a significant amount of their total energy from nonbiotic sources—that is, from something other than human and animal muscles fueled by the annual product of photosynthesis. If that turns out to be correct, future civilizations that learn to use energy sparingly may be able to accomplish some of the things that we currently do by throwing energy around with wild abandon, and they may also learn how to do remarkable things that are completely beyond our grasp today.

That's especially worth keeping in mind just now, because the decline and fall of industrial society is already happening. The insistence that this can't possibly happen may be popular, but again, history doesn't support it, and there are other issues involved as well. Comforting in the short term, the insistence that our civilization must last forever is a rich source of disaster and misery from any longer perspective, and the sooner each of us gets over it and starts to survey the wreckage around us, the better. Then we can make camp in the ruins, light a fire, get some soup heating in a salvaged iron pot, and begin to talk about where we can go from here.

Of course, not everyone is interested in that conversation. A great many people, for example, are interested only in answers that will allow them to keep on enjoying the absurd extravagance that passed, not too long ago, for an ordinary lifestyle among the industrial world's privileged classes and is becoming just a little bit less ordinary with every year that slips by. To such people I have nothing to say. Those lifestyles were possible only because the world's industrial nations burned through half a billion years of stored sunlight in a few short centuries and gave most of the benefits of that orgy of consumption to a relatively small fraction of their population; now that easily accessible reserves of fossil fuels are running short, the party's over.

I'm quite aware that that's a controversial statement. I field heated denunciations on a regular basis insisting that it just ain't so, that solar energy or nuclear fusion or perpetual motion or *something* will allow the industrial world's affluent classes to have their planet and eat it too. Printer's ink being unfashionable these days, a great many electrons have been inconvenienced on the internet to proclaim that this or that technology must surely allow the comfortable to remain comfortable, no matter what the laws of physics, geology, or economics have to say. It doesn't matter that the only alternative energy sources that have been able to stay in business even in a time of sky-high oil prices are those that can count on gargantuan government subsidies to pay their operating expenses; equally, the alternatives receive an even more gigantic "energy subsidy" from fossil fuels, which make them look much more economical than they otherwise would. Such reflections carry no weight with those whose sense of entitlement makes living with less unthinkable.

I'm glad to say that there are fair number of people who've gotten past that unproductive attitude, who have grasped the severity of the crisis of our time and are ready to accept unwelcome change in order to secure a livable future for our descendants. They want to know how we can pull modern civilization out of its current power dive and perpetuate it into the centuries ahead. I have no answers for them, either, because that's not an option at this stage of the game; we're long past the point at which decline and fall can be avoided, or even ameliorated on any large scale.

A decade ago, as noted in an earlier chapter, a team headed by Robert Hirsch and funded by the Department of Energy released a study outlining what would have to be done in order to transition away from fossil fuels before they transitioned away from us. What they found, to sketch out too briefly the findings of a long and carefully worded study, is that in order to avoid massive disruption, the transition would have to begin twenty years before conventional petroleum production reached its peak and began to decline.[10] There's a certain irony in the fact that 2005, the year this study was published, was also the year when conventional petroleum production peaked; the transition would thus have had to begin in 1985—right about the time, that is, that the Reagan administration in the US and its clones overseas were scrapping the promising steps toward just such a transition.

A grand transition that got under way in 2005, in other words, would have been too late, and given the political climate, it probably would have been too little as well. Even so, it would have been a much better outcome than the one we got, in which most people spent the past ten years insisting that we don't have to worry about depleting oilfields because fracking was going to save us all. At this point, thirty years after the point at which we would have had to get started, it's all very well to talk about some sort of grand transition to sustainability, but the time when that would have been possible came and went decades ago. We could have chosen that path, but we didn't, and the opportunities that might have been open to us then have long since gone whistling away with the wind.

Yet there's still an important set of opportunities open to us. It so happens that in very many cases, as already discussed, older, simpler, sturdier technologies work better, producing more satisfactory outcomes and fewer negative side effects, than their modern high-tech equivalents. Thus deliberate technological regression as public policy doesn't amount to a return to the caves—quite the contrary, it means a return to things that actually work. Given the range of confusions and ambiguities surrounding technological regression, it may be worth coining a new term for it. The one I have in mind is *retrovation*.

This is obviously backformed from "retro" + "innovation," but it's also "re-trove-ation," re-finding, rediscovery: an active process of searching through the many options the past provides, not a passive acceptance of some bygone time as a package deal. That's the strategy discussed in this book. The rhetoric of progress rejects that possibility, demands acquiescence to a very dubious logic that lumps "the past" together as a single thing, and insists that wanting any of it amounts to wanting all of it, with the worst features inevitably highlighted.

That latter insistence is central to the problems we face as a species. At the heart of most of the modern world's problems, as I've suggested, is the devout faith that human history is a straight line with no branches or meanders, leading onward and upward from the caves to the stars, and that every software upgrade, every new and improved product on the shelves, every lurch "forward"—however that conveniently floppy word happens to be defined from day to day by marketing flacks and politicians—therefore must lead toward that imaginary destination. That blind and increasingly untenable faith is the central reason why the only future different from the present that most people can imagine these days is either a rehash of the past in every detail or some kind of nightmare dystopia.

When you've driven down a blind alley and are sitting there with your bumper pressed against a brick wall, the only way forward starts by backing up. If you've been convinced by your society's core ideological commitments that "backing up" can only mean returning whole hog to the imaginary, awful past from which the ersatz messiah of progress is supposed to save us, you're stuck. There you sit, pushing uselessly on the pedal, hearing the engine labor and rattle, and watching the gas gauge move steadily toward that unwelcome letter E; it's no surprise that after a while, the idea of a street leading somewhere else starts to seem distinctly unreal.

Still, it remains the case that staying stuck against the brick wall leads nowhere useful. Those people who grasp this, shake themselves free of the dysfunctional ideology of progress, and explore other options are going to be the ones who make a difference in the shape the future will have on the far side of the crisis years ahead. Let go of the futile struggle to sustain the unsustainable, take the time and money and other resources that might be wasted in that cause, and do something less foredoomed with them: by embracing that strategy, there's a lot that can still be done, by means of retrovation and other relevant strategies, even in the confused and calamitous time that's breaking over us right now.

Notes

Preface

1. Greer 2015.
2. Greer 2016.

Chapter 1

1. Greer 2015.
2. Huebner 2005.
3. Daly 2014.
4. See Hardin 1968 and Ostrom 1990.
5. Dolan 1971 is the book on economics; Heinlein 1966 popularized the acronym.
6. Hornborg 2001.
7. Roberts 2013.
8. Scheyder 2015.
9. Turner and Alexander 2014.
10. See, for example, Rines 2015.
11. For a particularly egregious example, see Kelly 2010.
12. See the discussion of appropriate technology in Greer 2013.

Chapter 2

1. Seife 2008.
2. See, for example, Martin 2016.
3. Oreskes 2015.
4. See, for example, Catton 1980.
5. See Dimitrov 2007 for a useful analysis.
6. Prieto and Hall 2015.
7. Miller et al. 2011.
8. Pillar 2015.
9. László 1972.

10. Cited in Suskind 2004.
11. Illich 1974.
12. Parramore 2016.
13. Debord 1970.

Chapter 3

1. Butti and Perlin 1980.
2. Hirsch et al. 2005.
3. Coupland 2010.

Chapter 4

1. Chrisman 2013.
2. See, for example, Bonazzo 2015 and Hamilton 2016.
3. Moser 1998.
4. The show is archived at https://www.peakprosperity.com/podcast/92330/john-michael-greer-god-technological-progress-may-well-dead.
5. Kunstler 2005.
6. See Dewson 2015, among many others.

Chapter 5

1. Ward-Perkins 2005.
2. Greer 2009.
3. I have based much of the following on Allen 1978.

Chapter 6

1. Nathan and Mencken 1920.
2. See Riché 1976.
3. Siegelbaum 2014.
4. Yardeni et al. 2016.
5. Kennedy 1987.
6. Costanza et al. 1997.
7. Beiser 1962, pp. 26–27.
8. Korzybski 1994.
9. These points are discussed at length in Greer 2015.
10. Wolfram 2002, p. 23.
11. Wolfram 2002, p. 1.

Chapter 7

1. Trotman 2015 is a representative example.
2. Sumitra 2014.
3. *The Onion* 2015.
4. Parramore 2016.
5. Fleming 2016.
6. See, for example, Jacob 1997.
7. Roszak 1972, p. 64.
8. Jevons 1866.
9. Butti and Perlin 1980, pp. 72–4.
10. Weber 2003, p. 181.

Chapter 8

1. Howard 1935.
2. Louv 2005.
3. Toynbee 1939.
4. El Akkad 2015.
5. Nuccitelli 2013.
6. Korzybski 1994.
7. Greer 2008, pp. 225–240.
8. This is the central theme of Spengler 1926.
9. Greer 2009.
10. Hirsch et al. 2005.

Bibliography

Allen, Oliver E., *The Windjammers* (New York: Time-Life Books, 1978).

Beiser, Arthur, *The Earth* (New York: Time-Life Books, 1962).

Bonazzo, John, "Vox's Victorian Couple Is Twitter's Latest Punching Bag," *The Observer*, 9 September 2015, http://observer.com/2015/09/voxs-victorian-couple-is-twitters-latest-punching-bag/, accessed 22 November 2016.

Butti, Ken, and John Perlin, *A Golden Thread: 2500 Years of Solar Architecture and Technology* (Palo Alto, CA: Cheshire Books, 1980).

Catton, William R., Jr., *Overshoot: The Ecological Basis of Revolutionary Change* (Urbana, IL: University of Illinois Press, 1980).

Chrisman, Sarah A., *Victorian Secrets: What a Corset Taught Me about the Past, the Present, and Myself* (New York: Skyhorse Publishing, 2013).

Conkling, Philip, Richard Alley, Wallace Broecker, and George Denton, *The Fate of Greenland: Lessons from Abrupt Climate Change* (Cambridge, MA: Massachusetts Institute of Technology Press, 2011).

Costanza, R., R. d'Arge, R. de Groot, S. Farber, M. Grasso, B. Hannon, K. Limburg, A. Naeem, R. O'Neill, J. Paruelo, R. Raskin, P. Sutton, and M. van den Belt, "The value of the world's ecosystem services and natural capital," *Nature* 387 (1997), pp. 253–260.

Coupland, Douglas, "A radical pessimist's guide to the next 10 years," *The Globe and Mail*, 8 October 2010, theglobeandmail.com/news/national/a-radical-pessimists-guide-to-the-next-10-years/article1321040/?page=all, accessed 2 December 2016.

Daly, Herman, "Three Limits to Growth," 4 September 2014, steadystate.org/three-limits-to-growth/, accessed 2 December 2016.

Debord, Guy, *The Society of the Spectacle* (London: Black & Red, 1970).

Deffeyes, Kenneth *Hubbert's Peak* (Princeton, NJ: Princeton University Press, 2003).

Dewson, Andrew, "A disaster waiting to happen in Oklahoma? The link between fracking and earthquakes is causing alarm in an oil-rich town," *The Independent*, 6 April 2015, independent.co.uk/news/world/americas/a-disaster-waiting-to-happen-in-oklahoma-the-link-between-fracking-and-earthquakes-is-causing-alarm-10158524.html, accessed 2 December 2016.

Dimitrov, Krassen, "GreenFuel Technologies: A Case Study for Industrial Photosynthetic Energy Capture," March 2007, ecolo.org/documents/documents_in_english/biofuels-Algae-CaseStudy-09.pdf, accessed 16 October 2016.

Dolan, Edwin G., *TANSTAAFL: The Economic Strategy for Environmental Crisis* (New York: Holt, Rinehart and Winston, 1971).

Duncan, Richard C., "The life-expectancy of industrial civilization: the decline to global equilibrium," *Population and Environment* 14:4 (1993), pp. 325–357.

El Akkad, Omar, "Come hell or high water: The disaster scenario that is South Florida," *The Globe and Mail*, 17 July 2015, theglobeandmail.com/news/world/come-hell-or-high-water-the-disaster-scenario-that-is-south-florida/article25552300/, accessed 11 November 2016.

Ellis, John, *The Social History of the Machine Gun* (New York: Pantheon Books, 1975).

Fleming, David, *Lean Logic* (White River Junction, VT: Chelsea Green, 2016).

Freeman, Mae, and Ira Freeman, *You Will Go To The Moon* (New York: Random House, 1959).

Gibson, William, and Bruce Sterling, *The Difference Engine* (New York: Bantam Books, 1991).

Greer, John Michael, *The Long Descent: A User's Guide to the End of Industrial Civilization* (Gabriola Island, BC: New Society, 2008).

———, *The Ecotechnic Future: Envisioning a Post-Peak World* (Gabriola Island, BC: New Society, 2009).

———, *Apocalypse Not* (San Francisco: Viva Editions, 2011).

———, *Green Wizardry: Conservation, Solar Power, Organic Gardening, and Other Hands-On Skills from the Appropriate Technology Toolkit* (Gabriola Island, BC: New Society, 2013).

———, *After Progress: Reason and Religion at the End of the Industrial Age* (Gabriola Island, BC: New Society, 2015).

———, *Dark Age America* (Gabriola Island, BC: New Society, 2016).

———, *Retrotopia* (Danville, IL: Founders House, 2016).

Hamilton, Graeme, "Couple refused entry to Victoria's Butchart Gardens because they're too... Victorian," *The National Post*, 16 August 2016, news.nationalpost.com/news/doff-your-hat-or-you-cant-come-in-couple-refused-entry-to-victoria-garden-because-theyre-too-victorian, accessed 3 September 2016.

Hardin, Garrett, "The Tragedy of the Commons," *Science* Vol. 162 no. 3859 (1968), pp. 1243–1248.

Heinberg, Richard, *The Party's Over: Oil, War and the Fate of Industrial Societies* (Gabriola Island, BC: New Society, 2003).

———, *Peak Everything: Waking Up to the Century of Declines* (Gabriola Island, BC: New Society, 2007).

———, *Snake Oil: How Fracking's False Promise of Plenty Imperils Our Future* (Santa Rosa, CA: Post Carbon Institute, 2013).

Heinlein, Robert A., *The Moon is a Harsh Mistress* (New York: Putnam, 1966).

Herbert, Frank, *Dune* (New York: Ace Books, 1965).

Hesse, Hermann, *The Glass Bead Game*, trans. Richard and Clara Winston (New York: Holt, Rinehard, & Winston, 1969).

Hirsch, Robert L., Roger Bezdek, and Robert Wendling, *Peaking of World Oil Production: Impacts, Mitigation, and Risk Management* (Washington, DC: US Department of Energy, 2005).

Hornborg, Alf, *The Power of the Machine: Global Inequalities of Economy, Technology, and Environment* (San Francisco: Alta Mira Press, 2001).

Howard, Robert E., "Beyond the Black River," *Weird Tales* 25 (1935), nos. 5 and 6.

Huebner, Jonathan, "A possible declining trend for worldwide innovation," *Technological Forecasting & Social Change* 72 (2005), pp. 980–986.

Illich, Ivan, *Energy and Equity* (New York: Harper & Row, 1974).

Jacob, Margaret C., *Scientific Culture and the Making of the Industrial West* (New York: Oxford University Press, 1997).

Jevons, William Stanley, *The Coal Question* (London: Macmillan, 1866).

Kelly, Kevin, *What Technology Wants* (New York: Viking, 2010).

Kennedy, Paul, *The Rise and Fall of Great Powers* (New York: Random House, 1987).

Korzybski, Alfred, *Science and Sanity: An Introduction to Non-Aristotelian Systems and General Semantics* (New York: Institute of General Semantics, 1994).

Kunstler, James Howard, *The Long Emergency* (New York: Atlantic Monthly Press, 2005).

———, "The Psychology of Previous Investment," *Raise the Hammer*, 21 October 2005, raisethehammer.org/article/181, accessed 2 December 2016.

Kurzweil, Ray, *The Singularity is Near* (New York: Viking, 2005).

Landis, Frank, *Hot Earth Dreams* (North Charleston, SC: Createspace, 2016).

László, Ervin, *Introduction to Systems Philosophy* (New York: Gordon and Breach, 1972).

Louv, Richard, *Last Child in the Woods: Saving Our Children from Nature-Deficit Disorder* (Chapel Hill, NC: Algonquin Books, 2005).

Lynas, Mark, *Six Degrees: Our Future on a Hotter Planet* (Washington DC: National Geographic, 2008).

Martin, Richard, "Why the world's largest nuclear fusion project may never succeed," *MIT Technology Review*, 4 May 2016, technologyreview.com/s/601388/why-the-worlds-largest-nuclear-fusion-project-may-never-succeed/, accessed 2 December 2016.

Meadows, Donnella, Dennis Meadows, Jórgen Randers, and William W. Behrens III, *The Limits to Growth* (New York: Universe, 1972).

Meyer, Ali, "Labor force participation has hovered near a 37-year low for 11 months," *CNS News*, m.cnsnews.com/news/article/ali-meyer/628-labor-force-participation-has-hovered-near-37-year-low-11-months, accessed 9 July 2016.

Miller, L. M., F. Gans, and A. Kleidon, "Estimating maximum global land surface wind power extractability and associated climatic consequences," *Earth Systems Dynamics*, 11 February 2011, earth-syst-dynam.net/2/1/2011/esd-2-1-2011.html, accessed 2 December 2016.

Moser, Stephanie, *Ancestral Images: The Iconography of Human Origins* (Ithaca, NY: Cornell University Press, 1998).

Mumford, Lewis, *Technics and Civilization* (New York: Harcourt, Brace & Co., 1934).

Nathan, George Jean, and H. L. Mencken, *The American Credo* (New York: A. A. Knopf, 1920).

Nuccitelli, Dana, "A grand solar minimum would barely make a dent in human-caused global warming," *The Guardian*, 14 August 2013, theguardian.com/environment/climate-consensus-97-per-cent

/2013/aug/14/global-warming-solar-minimum-barely-dent, accessed 30 November 2016.

The Onion, "Health Experts Recommend Standing Up At Desk, Leaving Office, Never Coming Back," 6 February 2015, theonion.com /article/health-experts-recommend-standing-up-at-desk-leavi -37957, accessed 2 December 2016.

Ophuls, William, *Immoderate Greatness: Why Civilizations Fail* (North Charleston, NC: CreateSpace, 2012).

Oreskes, Naomi, "There is a new form of climate denialism to look out for—so don't celebrate yet," *The Guardian*, 16 December 2015, theguardian.com/commentisfree/2015/dec/16/new-form-climate -denialism-dont-celebrate-yet-cop-21, accessed 4 April 2016.

Oskin, Becky, "Catastrophic Collapse of West Antarctic Ice Sheet Begins," *LiveScience*, 12 May 2014, livescience.com/45534-west -antarctica-collapse-starts.html, accessed 25 December 2015.

Ostrom, Elinor, *Governing the Commons: The Evolution of Institutions for Collective Action* (Cambridge, UK: Cambridge University Press, 1990).

Parramore, Lynn, "Before capitalism, medieval peasants got more vacation time than you. Here's why," *Evonomics*, 2 November 2016, evonomics.com/capitalism-medieval-peasants-got-vacation-time -heres/, accessed 2 December 2016.

Pillar, Paul R., "The Odd American View of Negotiation," *The National Interest*, 20 June, 2015, nationalinterest.org/blog/paul-pillar/the -odd-american-view-negotiation-13159, accessed 23 September 2016.

Prieto, Pedro, and Charles A. S. Hall, *Spain's Photovoltaic Revolution: The Energy Return on Investment* (New York: Springer Publishing, 2015).

Reighley, Kurt B., *The United States of Americana* (New York: Harper, 2010).

Riché, Pierre, *Education and Culture in the Barbarian West* (Columbia: University of South Carolina Press, 1976).

Rines, Samuel, "Secular Stagnation: The Dismal Fate of the Global Economy?" *The National Interest*, 11 September 2015, nationalinterest .org/feature/secular-stagnation-the-dismal-fate-the-global-economy -13819, accessed 23 December 2015.

Roberts, David, "None of the world's top industries would be profitable if they paid for the natural capital they use," 17 April, 2013, *Grist*,

grist.org/business-technology/none-of-the-worlds-top-industries-would-be-profitable-if-they-paid-for-the-natural-capital-they-use/, accessed 5 December 2015.

Robinson, Kim Stanley, *The Years of Rice and Salt* (New York: Bantam, 2002).

Roszak, Theodore, *Where the Wasteland Ends* (New York: Harper & Row, 1962).

Scheyder, Ernest, "North Dakota: oil producers aim to cut radioactive waste bills," *Reuters*, 28 January, 2015, reuters.com/article/2015/01/28/us-usa-north-dakota-waste-idUSKBN0L11Z420150128, accessed 5 December 2015.

Seife, Charles, *Sun in a Bottle: The Strange History of Fusion and the Science of Wishful Thinking* (New York: Penguin, 2008).

Siegelbaum, Debbie, "Project Hieroglyph: Fighting society's dystopian future," BBC, 4 September 2014, bbc.com/news/magazine-28974943, accessed 4 February 2015.

Snyder, Timothy, *Bloodlands: Europe between Hitler and Stalin* (New York: Basic Books, 2010).

Spengler, Oswald, *The Decline of the West, Vol. 1: Form and Actuality*, trans. Charles Francis Atkinson (New York: Alfred A Knopf, 1926).

Sumitra, "Living Life Like in the 1950s—America's Rockabilly Community," *Oddity Central*, 24 January 2014, odditycentral.com/pics/living-life-like-in-the-1950s-americas-rockabilly-community.html, accessed 21 November 2016.

Suskind, Ron, "Faith, certainty, and the presidency of George W. Bush," *New York Times Magazine*, 17 October 2004, nytimes.com/2004/10/17/magazine/faith-certainty-and-the-presidency-of-george-w-bush.html?_r=0, accessed 2 December 2016.

Tolkien, J. R. R., *The Fellowship of the Ring* (New York: Ballantine Books, 1965).

Toynbee, Arnold, *A Study of History, Vol. IV, The Breakdowns of Civilizations* (London: Oxford University Press, 1939).

Trotman, Andrew, "Kindle sales have 'disappeared', says UK's largest book retailer," *The Telegraph*, 6 January 2015, telegraph.co.uk/finance/newsbysector/retailandconsumer/ 11328570/Kindle-sales-have-disappeared-says-UKs-largest-book-retailer.html, accessed 18 January 2015.

Tuchman, Barbara, *The Proud Tower* (New York: Macmillan, 1966).

Turner, Graham, and Cathy Alexander, "Limits to Growth was right. New research shows we're nearing collapse," *The Guardian*, 2 September, 2014, theguardian.com/commentisfree/2014/sep/02/limits-to-growth-was-right-new-research-shows-were-nearing-collapse, accessed 5 December 2015.

Ward-Perkins, Bryan, *The Fall of Rome and the End of Civilization* (Oxford: Oxford University Press, 2005).

Weaver, Andrew J., et al., "Meltwater Pulse 1A from Antarctica as a Trigger of the Alleröd-Bölling Warm Interval," *Science* vol. 299 no. 5613 (14 March 2003), pp. 1709–13.

Weber, Max, *The Protestant Ethic and the Spirit of Capitalism*, trans. Talcott Parsons (Minneola, NY: Dover Books, 2003).

Wolfram, Steven, *A New Kind of Science* (Champaign, IL: Wolfram Media, 2002).

Yardeni, Edward, Debbie Johnson, and Mali Quintana, "Standard of Living: Income Shares by Quintiles," *Yardeni Research, Inc.*, 14 September 2016, yardeni.com/pub/standardlivequintiles.pdf, accessed 10 October 2016.

Index